AF179755

Aluminium Alloys - Design and Development of Innovative Alloys, Manufacturing Processes and Applications

Edited by Giulio Timelli

Published in London, United Kingdom

IntechOpen

Supporting open minds since 2005

Aluminium Alloys - Design and Development of Innovative Alloys, Manufacturing Processes and Applications
http://dx.doi.org/10.5772/intechopen.95155
Edited by Giulio Timelli

Contributors
Lallia Belkacem, Elisa Fracchia, Mario Rosso, Ahmet Atak, Derya Dispinar, Kerem Can Dizdar, Hayati Sahin, Furkan Tezer, Samarendra Roy, Shibayan Roy, Rocio Maricela Ochoa Palacios, Citlaly Castillo Rodriguez, Jesus Torres Torres, Alfredo Flores Valdes, Perla Janet Resendiz Hernandez, Toshio Haga, Nur Ayuni Jamal, Farazila Yusof, Yusilawati Ahmad, Norhuda Hidayah Nordin, Suraya Sulaiman, Bhanu Prakash Palampalle, Babu Dharmalingam, Devika Royal, Aishah Najiah Dahnel, Muhammad Azim Mirza Mohd Farid, Mohamad Noor Ikhwan Naiman, Ahmad Faris Abdul Rahman, Nur Munirah Meera Mydin

© The Editor(s) and the Author(s) 2022
The rights of the editor(s) and the author(s) have been asserted in accordance with the Copyright, Designs and Patents Act 1988. All rights to the book as a whole are reserved by INTECHOPEN LIMITED. The book as a whole (compilation) cannot be reproduced, distributed or used for commercial or non-commercial purposes without INTECHOPEN LIMITED's written permission. Enquiries concerning the use of the book should be directed to INTECHOPEN LIMITED rights and permissions department (permissions@intechopen.com).
Violations are liable to prosecution under the governing Copyright Law.

Individual chapters of this publication are distributed under the terms of the Creative Commons Attribution 3.0 Unported License which permits commercial use, distribution and reproduction of the individual chapters, provided the original author(s) and source publication are appropriately acknowledged. If so indicated, certain images may not be included under the Creative Commons license. In such cases users will need to obtain permission from the license holder to reproduce the material. More details and guidelines concerning content reuse and adaptation can be found at http://www.intechopen.com/copyright-policy.html.

Notice
Statements and opinions expressed in the chapters are these of the individual contributors and not necessarily those of the editors or publisher. No responsibility is accepted for the accuracy of information contained in the published chapters. The publisher assumes no responsibility for any damage or injury to persons or property arising out of the use of any materials, instructions, methods or ideas contained in the book.

First published in London, United Kingdom, 2022 by IntechOpen
IntechOpen is the global imprint of INTECHOPEN LIMITED, registered in England and Wales, registration number: 11086078, 5 Princes Gate Court, London, SW7 2QJ, United Kingdom
Printed in Croatia

British Library Cataloguing-in-Publication Data
A catalogue record for this book is available from the British Library

Additional hard and PDF copies can be obtained from orders@intechopen.com

Aluminium Alloys - Design and Development of Innovative Alloys, Manufacturing Processes and Applications
Edited by Giulio Timelli
p. cm.
Print ISBN 978-1-83968-729-7
Online ISBN 978-1-83968-730-3
eBook (PDF) ISBN 978-1-83968-731-0

For EU product safety concerns:
IN TECH d.o.o., Prolaz Marije Krucifikse Kozulić 3, 51000 Rijeka, Croatia, info@intechopen.com or visit our website at intechopen.com.

We are IntechOpen, the world's leading publisher of Open Access books

Built by scientists, for scientists

5,900+
Open access books available

144,000+
International authors and editors

180M+
Downloads

156
Countries delivered to

Our authors are among the
Top 1%
most cited scientists

12.2%
Contributors from top 500 universities

CLARIVATE ANALYTICS BOOK CITATION INDEX INDEXED

WEB OF SCIENCE™

Selection of our books indexed in the Book Citation Index (BKCI) in Web of Science Core Collection™

Interested in publishing with us?
Contact book.department@intechopen.com

Numbers displayed above are based on latest data collected.
For more information visit www.intechopen.com

Meet the editor

Giulio Timelli is an Associate Professor of Metallurgy and Innovative Metallurgical Processes, at the University of Padua, Italy. After obtaining an MS in Industrial Engineering and a Ph.D., he was a visiting scientist at Jonkoping University, Sweden. In 2006, he was a researcher at the Norwegian University of Science and Technology. In 2008 and 2011, he won the "A. Daccò Prize" for the best Italian papers on solidification topics. He is the scientific director of several industrial research programs in collaboration with different local and international companies. He is vice president of the Diecasting Technical Group and president of the Light Metals Technical Group of the Italian Association of Metallurgy (AIM). He is a member of the CEN Technical Committee for the development of European standards for aluminium alloys. His research interests include the solidification of Al and Mg alloys, the development of light alloys' processes, the characterization of Al matrix composites, the development of heat treatment processes for light alloys, and the numerical simulation of foundry and wrought processes.

Contents

Preface

During the last decades, there has been a significant increase in the use of light alloys, which have the potential to replace ferrous alloys in engineering sectors. In general, aluminium alloys dominate over other light metals thanks to good properties, including their excellent compromise between mechanical properties and lightness, corrosion resistance, easy recyclability, and relatively low cost. The global demand for aluminium is growing and approximately 63 million tons of aluminium alloys were produced in 2018. In Europe alone, 7.8 million tons of aluminium alloys were produced in the same period. This increased production was possible thanks to the development of innovative manufacturing processes and new generations of aluminium alloys able to increase the final mechanical performance and the processability of the material. Some of these processes include casting, drawing, forging, rolling, extrusion, welding, powder metallurgy, and additive manufacturing.

This book provides a theoretical and practical understanding of the metallurgical principles in the design and development of innovative aluminium alloys and composites. It also presents and discusses a new generation of manufacturing processes and applications. Furthermore, the book is heavily illustrated to make the concepts clear.

The editor of this book hopes that practicing engineers, technicians, students, and all those interested in aluminium alloys, in particular the technical staff of aerospace, automotive, and defence industries, will find it useful. The editor would like to acknowledge the contributing authors for their excellent chapters.

Giulio Timelli
Department of Management and Engineering,
University of Padua,
Vicenza, Italy

Section 1

Design and Development of Innovative Aluminium Alloys

Chapter 1

Development and Characterization of New Functionally Graded Aluminium Alloys

Elisa Fracchia and Mario Rosso

Abstract

Nowadays, aluminium alloys are adopted mainly to produce engineering and automotive components. The present investigation aims to design, cast and characterize novel functionally graded materials (FGMs) produced using Al-Mg and Al-Si alloys by gravity casting technique. Alloys were sequentially cast into a mould to obtain an FGM to realizing great mechanical and metallurgical bonding. Zn addition was further performed in FGM to increase the mechanical properties, thanks to the nucleation of the intermetallic phases $MgZn_2$. Castings were subsequently mechanically tested by tensile tests, bending tests, hardness and microhardness measures to assess the products' quality. Microstructural characterizations were performed along the FGM to assess the metallurgical bonding and evaluate the microstructures obtained. Fracture, microstructural and compositional analysis will highlight the quality of this new FGM proposed. Possible applications of these materials are suggested, as automotive pistons or structural components.

Keywords: FGM, Al-Si alloy, Al-Mg alloy, tensile tests, bending test, microstructures, SEM analysis

1. Introduction

The last few decades were characterized by significant growth of the aluminium market. This growing interest in aluminium alloys is attributable to their excellent properties such as good or excellent specific mechanical strength, lightweight and generally good corrosion resistance [1]. During previous years, in the automotive market, these properties have allowed replacing some heavy components made in steel with the same components made in aluminium, leading to cost savings, structural lightening and CO_2 emission reduction. However, if, on the one hand, aluminium alloys have good properties that permit the production of components with homogenous characteristics providing high performances, on the other hand, some applications may require specific features as graded structures. Functionally graded materials (FGMs) have graded structures characterized by different compositions or microstructures, as deeply investigated in [2]. In particular, gradient types were classified as chemical composition gradients, porosity gradients or microstructural gradients [3], while microstructural gradients are further subdivided

 IntechOpen

into fraction gradient, shape gradient, orientation gradient and size gradient [4]. Various production methods such as powder metallurgy, centrifugal casting, vapour deposition [5] or additive manufacturing [6] can be used to produce FGMs. FGMs usage involves various markets, such as biomedical, chemical, aerospace, electronics, nuclear or energy; for instance, in [7], the powder metallurgy method was used, and the authors successfully fabricated a porosity gradient Ti-Zr FGM for biomedical application.

As aluminium-FGMs regard, various publications concern the functionally graded metal-ceramic composites rather than metal-metal ones. In [8], authors realized a layered Al-SiC FGM by powder metallurgy technique. Each layer was composed of different SiC content, and the consolidation between each layer was assured by cold compaction. Similarly, in [9] a layered Al-Al_2O_3 FGM was suggested. Other manuscripts suggest the adoption of Al alloys instead of the commercially pure aluminium powder. In particular, in [10], FGM produced with aluminium alloy A7075 was suggested for gears or brake drums. In addition to powder metallurgy, FGMs may be prepared in other ways: in [11], authors proposed the centrifugal casting technique to realize A356-SiC FGMs. The mould rotation permits a radial distribution of SiC particles, and in the end, the samples could be distinguished into three different zones—reinforced zone (outer zone), transitional zone (middle zone) and the unreinforced alloy (inner zone). Centrifugal casting was also successfully adopted in [12] to produce automotive pistons by adopting two aluminium alloys—A336 and A242. As centrifugal casting regards, different production parameters must be considered, as pointed up in the review article [13].

On the other hand, gravity casting is a newish casting method used to produce FGMs. In [14], the authors used gravity casting to realize two metal-metal FGM, adopting the aluminium alloys A390-A319 (both casting alloys) and A390-A6061 (casting and wrought alloys). In particular, to obtain a good metallurgical bonding between alloys, some casting parameters were carefully monitored, as the time gap during sequential pouring and the pouring temperature. In this work, metal-metal FGMs were realized by controlling the mould filling using gravity casting [15]. Alloys EN AC 47000 and EN AC 51100 were used, with and without Zn addition, to assess the mechanical properties and the metallurgical bonding in this new kind of FGM. FGMs properties were then compared to the properties of the single alloys.

2. Materials and methods

2.1 Materials

Aluminium alloys adopted to realize FGMs are EN AC 47000 and EN AC 51100. Details about alloys chemical composition are reported in **Table 1**. Alloy EN AC 51100 (AlMg3) has a wide range of applications in the automotive industry; Mg increases corrosion resistance and solution hardening [16].

EN AC 47000 (AlSi12(Cu)) is a eutectic alloy having Si as the primary alloying element. Si and Cu increase mechanical properties, in particular after heat treatment.

Commercially pure Zn ingot was used as an alloying element in a quantity of 1.5 w.t.% in some specific samples.

2.2 FGM production

FGMs were realized by controlling the mould filling during the gravity casting [17, 18]. The alloys are melted in two different crucibles while the mould is painted

EN AC 47000							
w.t.%	Cu	Mg	Si	Fe	Mn	Zn	Al
	0.62	0.09	12.98	0.54	0.26	0.47	Bal.
EN AC 51100							
w.t.%	Cu	Mg	Si	Fe	Mn	Zn	Al
	0.042	2.786	0.432	0.285	0.188	0.049	Bal.

Table 1.
FGM alloys composition.

with BN-based stop off-paint and preheated at 400°C. The mould is a C40 steel mould having dimensions 85 mm (depth) × 200 mm (height) × 15 mm (width). The alloys were poured in a precise amount to fill half of the mould figures (see dotted lines in **Figure 1A** and **B**). Approximately 150 g of EN AC 51100 alloy was poured at 710°C into the mould, as indicated by the orange arrow in **Figure 1A**, filling half of it. After the first casting, 185 g of alloy EN AC 47000 was poured, at 710°C, over EN AC 51100 to fill the second half of the mould, as shown in **Figure 1B**. The pouring of alloy EN AC 47000 was performed without any time delay after casting the first composition. Castings are manually extracted from the mould in the fastest way possible by using pullers and then quenched in water at 25°C; the manual extraction method may cause little change in cooling rate (that was not specifically measured) between subsequent castings. Because of the mould shape, there is no specific pouring channel for the second alloy, EN AC 47000, which must be poured into two different parts of the mould, as indicated in **Figure 1B**. This mould configuration may affect the junction tightness inside castings, causing the presence of gas porosities

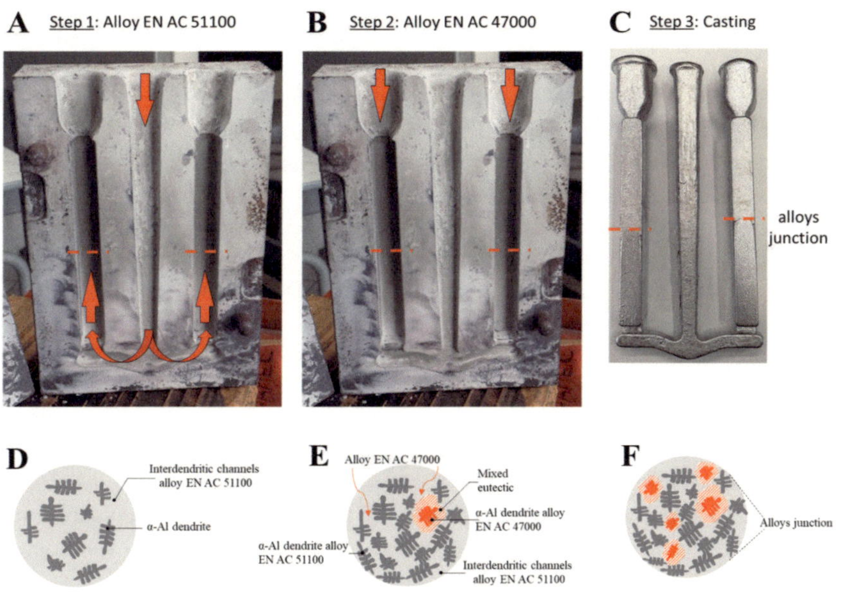

Figure 1.
Filling operations. A: Step 1, half-mould image; the casting of alloy EN AC 51000 inside the mould carried out as indicated by the orange arrow. B: Step 2, casting alloy EN AC 47000 inside the mould carried out as indicated by orange arrows. C: Step 3, casting obtained. D: Step 1 of casting, nucleation of α-Al dendrites near nucleation site and mould walls after the casting of composition EN AC 51100. E: Step 2 of casting, nucleation of α-Al dendrites of alloy EN AC 47000 into the interdendritic channels of alloy EN AC 51100 previous poured. F: EN AC 47000 alloy into the interdendritic channels of alloy EN AC 51100 after the pouring.

Castings	Alloys	Casting details
#1	FGM. EN AC 51100 and EN AC 47000	—
#2	FGM. EN AC 51100 and EN AC 47000 + 1.5w.t.% Zn	1.5w.t.% Zn in alloy EN AC 47000 (~ 2.8 g)
#3	FGM. EN AC 51100 + 1.5w.t.% Zn and EN AC 47000 + 1.5w.t.% Zn	Zn in EN AC 47000 (~ 2.8 g) and in EN AC 51100 (~ 2.2 g)
#4	Single alloy EN AC 47000 + 1.5w.t.% Zn	Zn addition (~ 2.8 g)
#5	Single alloy EN AC 51100 + 1.5w.t.% Zn	Zn addition (~ 2.2 g)
#6	Single alloy EN AC 51100	—
#7	Single alloy EN AC 47000	—

Table 2.
Experimental conditions.

and oxides. An example of the casting obtained is shown in **Figure 1C**. Each FGM bar measures 25 mm (depth) × 125 mm (height) × 15 mm (width), and the interface between alloys is approximately placed in the middle.

As mechanical properties are relevant in automotive FGM, Zn addition was performed to increase mechanical strength. For this reason, three different types of FGM were realized—FGM without Zn, FGM with Zn addition in alloy EN AC 47000 and finally FGM with Zn addition in both the alloys. Moreover, in order to evaluate the alloys' mechanical properties, single-alloy specimens have been cast, with and without Zn addition. The different experimental conditions are summarized in **Table 2**.

2.3 Mechanical and microstructural tests

With the aim to evaluate the mechanical properties of the produced FGMs and compare them with those of the single alloys, tensile tests, three-point bending tests and microhardness measures were made. Tensile tests were performed following the norm ASTM B557–15. Specimens were machined in a plate dog-bone shape from the rectangular castings; in FGM tensile specimens, the interface between the alloys was placed in the middle of the samples. Since the properties of the FGM without Zn addition are the work's focus, a higher number of the tensile test specimens were produced. Overall, these were realized six specimens for this type of FGMs without Zn, while the other types of casting were machined with two samples.

Three-point bending tests were performed on bar measures 10 mm × 10 mm × 60 mm, adopting a support span of 40 mm, test speed 0.004 1/s and a preload of 5 N. During the tests, the load is applied in the middle of the specimens, while on the opposite side the sample is supported by two wedges. In FGM bending specimens, the interface between the alloys was placed in the middle of the samples. Overall, there were six specimens for FGMs without Zn, while the other FGMs were machined in two samples. Four specimens were tested for the single alloys (castings with and without Zn).

Microhardness Vickers tests were performed on each specimen. FGMs interfaces were subjected to a microhardness matrix 8 × 8; the distance between each indentation was 150 μm, the applied load was 15 gf for 15 s and the diagonal was measured after the indentation varies almost from 25 to 35 μm. Single alloys were also tested, performing five indentations for each sample. Microhardness was chosen instead of Vicker hardness because it may be more sensitive to the slight hardness variations near the interface of the FGM samples. In fact, the junction areas of alloys are pretty low, of a few hundred microns.

Microstructures were evaluated through SEM microscope, and EDS semiquantitative analyses were also carried out. Castings were cut and polished by using SiC papers from 180 to 2500 grit, and then, colloidal silica having a granulometry of 0.03 μm was used for mirror-polishing. Finally, specimens were etched with Keller's reagent. Fracture surfaces after tensile tests were also investigated through SEM analysis.

3. Results and discussion

3.1 Intermetallic phases

Figure 2 shows microstructures obtained at SEM microscope for specimens of alloy EN AC 51100 with Zn and EN AC 51100 without Zn addition. Different intermetallic phases are noticeable from **Figure 2a**: Al-Fe-Mn-Si phases (spectrums 1 and 4), the eutectic α-Al/Al_3Mg_2 phase (spectrum 5) and Mg_2Si phase (spectrums 2 and 5). Mainly, Al_3Mg_2 is characterized by the typical acicular shape and the black colour, as in spectrum 2 [16]. Mg_2Si can be present inside the eutectic α-Al/Al_3Mg_2 phase (grey precipitations). Similar results were noticed in **Figure 2b** for the analysed specimen of alloy EN AC 51100 with Zn addition. In particular, Zn was detected in spectrums 3, 4 and 5, along with Mg and Si. White intermetallic

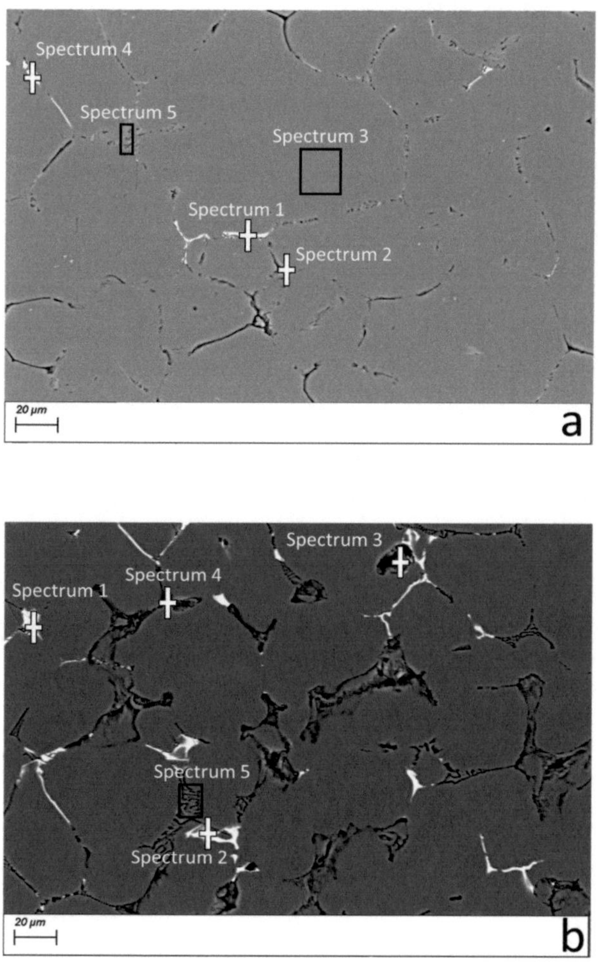

(a)

Spectrum 1

Elements	Wt%
Al	70.6
Fe	11.9
Si	8.9
Mn	7.8
Cu	0.4
Mg	0.4

Spectrum 2

Elements	Wt%
Al	78.7
Si	20.8
Mg	0.5

Spectrum 3

Elements	Wt%
Al	98.1
Mg	1.2

Spectrum 4

Elements	Wt%
Al	73
Fe	10.9
Si	8.1
Mn	8.0

Spectrum 5

Elements	Wt%
Al	77.0
Si	20.3
Mg	2.7

(b)

Spectrum 1

Elements	Wt%
Al	70.6
Fe	19.1
Si	6.1
Mn	3.8
Mg	0.5

Spectrum 2

Elements	Wt%
Al	66.3
Fe	22.2
Si	6.7
Mn	4.9

Spectrum 3

Elements	Wt%
Al	95.8
Mg	2.8
Zn	0.9
Si	0.4

Spectrum 4

Elements	Wt%
Al	90.9
Si	4.0
Mg	3.8
Zn	1.2

Spectrum 5

Elements	Wt%
Al	90.5
Si	4.8
Mg	4.0
Zn	0.8

Figure 2.
SEM-EDS analysis of specimen EN AC 51100 (a) and EN AC 51100 + Zn (b).

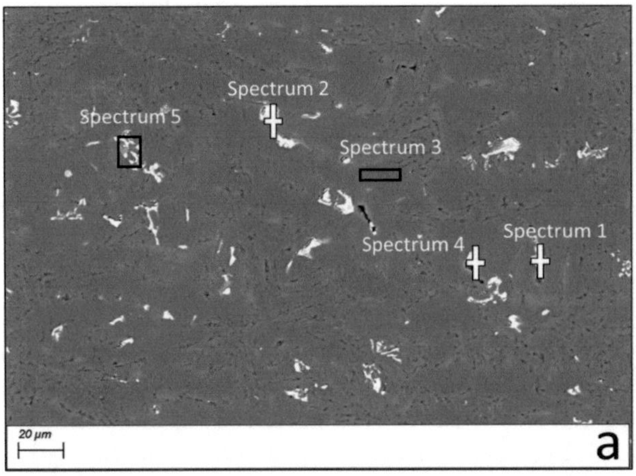

Spectrum 1

Elements	Wt%
Al	60.0
Cu	15.3
Fe	11.0
Si	9.3
Mn	2.7
Mg	1.8

Spectrum 2

Elements	Wt%
Al	59.9
Fe	19.5
Si	9.9
Mn	9.4
Cu	1.2

Spectrum 3

Elements	Wt%
Al	66.4
Si	33.6

Spectrum 4

Elements	Wt%
Al	54.2
Cu	28.2
Si	10.6
Mg	7.0

Spectrum 5

Elements	Wt%
Al	77.3
Fe	8.0
Si	7.8
Mn	3.5
Mg	1.8
Cu	1.6

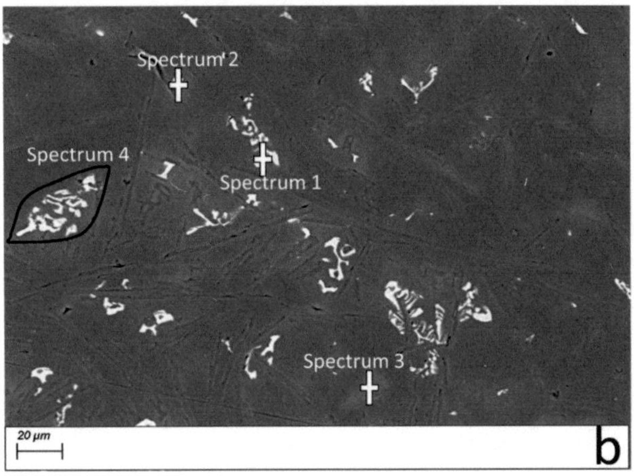

Spectrum 1

Elements	Wt%
Al	62.9
Fe	17.5
Si	9.2
Mn	8.2
Zn	1.2
Cu	1.0

Spectrum 2

Elements	Wt%
Al	58.3
Cu	32.0
Si	4.5
Mg	3.1
Zn	2.0

Spectrum 3

Elements	Wt%
Si	99.2
Al	0.8

Spectrum 4

Elements	Wt%
Al	82.6
Si	7.5
Fe	2.9
Cu	2.8
Zn	2.8
Mn	1.5

Figure 3.
SEM-EDS analysis of specimen EN AC 47000 (a) and EN AC 47000 + Zn (b).

compounds contain a high amount of Fe and Mn were also noticed in **Figure 2a** spectrum 4 and **Figure 2b** spectrum 2, identifiable as α-Al (Fe, Mn) Si phase.

Figure 3 shows microstructures obtained at SEM microscope for specimens of alloy EN AC 47000 with Zn and EN AC 47000 without Zn addition. In **Figure 3a**, white intermetallic compounds contain Fe, Mn and Cu, (spectrums 1, 2 and 5) similar to those observed in alloy EN AC 51100 were observed. Spectrum 4 is an intermetallic phase Al-Si-Mg-Cu also known as Q phase [19]. Similarly, intermetallic phases detected in **Figure 3b** highlight the presence of the Q phase in spectrum 2. Spectrum 3 highlights a eutectic silicon polygonal particle. The Zn was noticed in the intermetallic Fe-based, commonly known as Chinese script, α-AlFeSiCuMg.

3.2 FGMs interfaces

Figure 4 reports the SEM-EDS maps of the FGMs interfaces obtained. **Figure 4a** highlights the Si diffusion into Mg-based alloy, especially into the eutectic regions α-Al/Al_3Mg_2.

Iron-based intermetallic compounds are largely diffused into EN AC 47000 alloy, as noticeable from the maps; furthermore, Mg was detected in both the alloys with a slight depletion at the interface. Although Si was detected at the interface

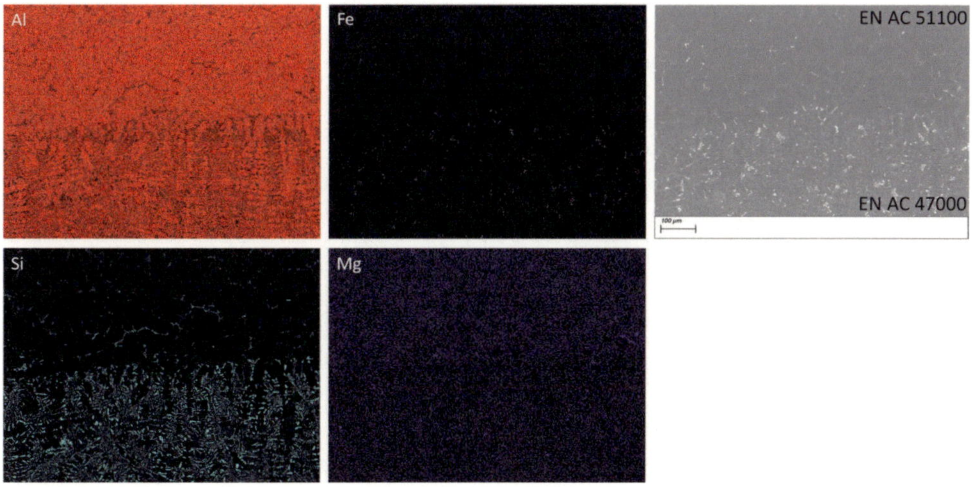

Figure 4.
EDS maps analysis in FGM without Zn interface.

into the EN AC 51100 bulk, in the alloy bulk Si was noticed only in a few intermetallic phases. In fact, EDS map analysis conducted into the EN AC 51100 bulk did not evidence appreciable amounts of silicon, as is possible to note in **Figure 5**.

After Zn addition, Zn-based intermetallic compounds are detected. **Figure 6** shows the SEM-EDS maps for FGM with Zn addition in alloy EN AC 47000. Overall, it seems that the Zn amount was not enough to be appreciated in the EDS maps near the interface; on the other hand, Zn was clearly noticed in the bulk alloy of the sample made in single alloy EN AC 47000 + Zn, as is shown in **Figure 7**.

Interface maps for FGM containing Zn in both alloys clearly show the presence of Zn, with a slight depletion into the interface between the compositions (see **Figure 8**). Furthermore, as also noticed in the other FGMs, Mg depletion was observed near the alloy interface. This depletion may be explained considering the alloy mixing that took place during the pouring of the second composition; in fact, near the interface, these were mainly noticed iron-based intermetallic phases (**Figures 4** and **6**).

No defects such as oxide layers or shrinkage were observed in the FGM interfaces.

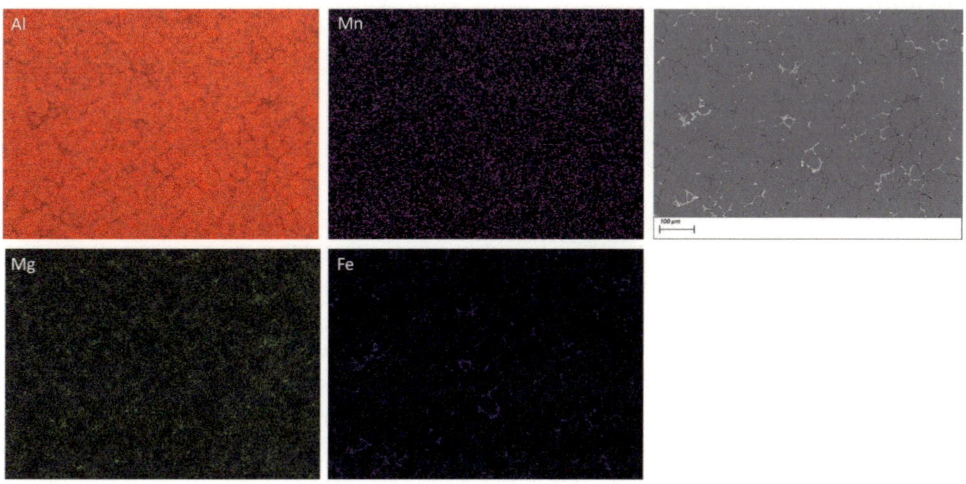

Figure 5.
EDS maps analysis of EN AC 51100 bulk.

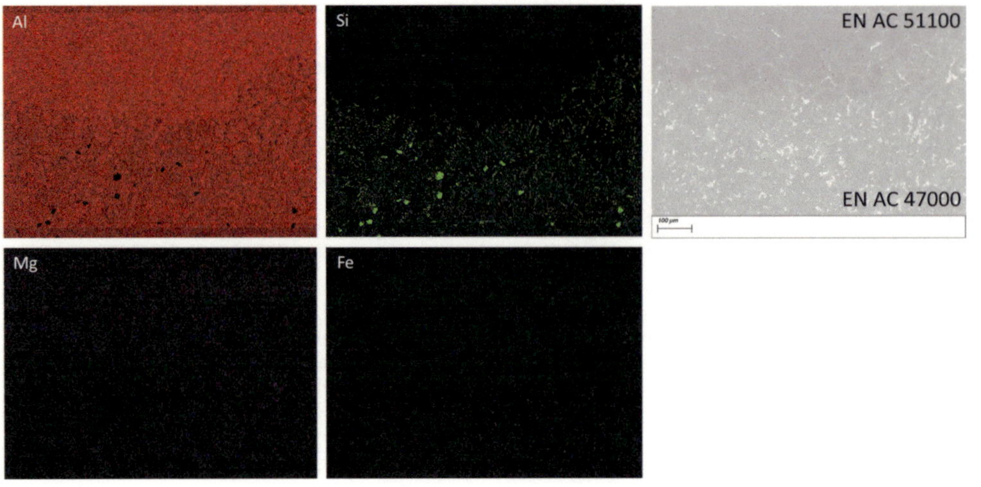

Figure 6.
EDS maps analysis in FGM (with Zn addition in alloy EN AC 47000) interface.

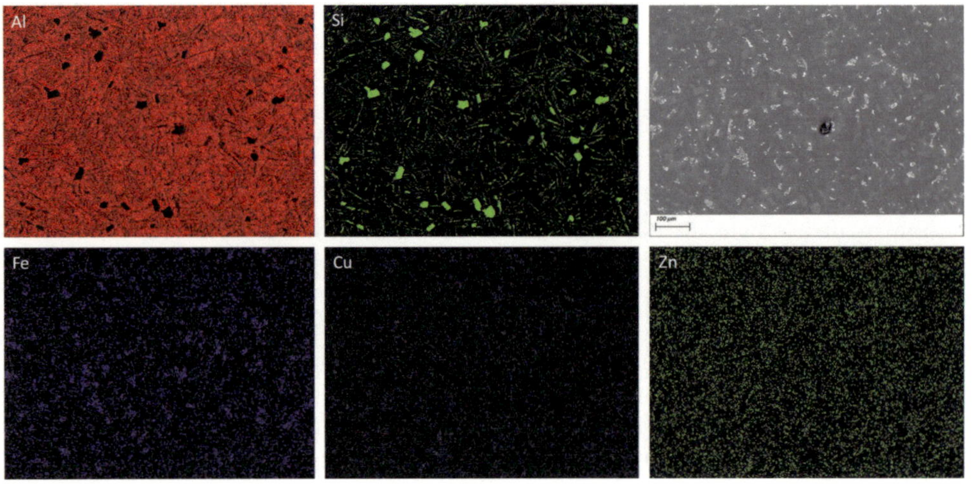

Figure 7.
EDS maps analysis of EN AC 47000 bulk.

3.3 Mechanical tests

3.3.1 Microhardness

Microhardness measured in alloys with and without Zn additions is shown in the bar chart of **Figure 9**. As FGMs regard, the microhardness matrixes measured through the interfaces are shown in **Figure 10**.

Average interface hardness appears similar in FGM and in FGM with Zn addition in alloy EN AC 47000, respectively 72 ± 16 HV0.15 and 69 ± 15. On the other hand, in the FGM with Zn addition in both alloys, the interface was characterized by a higher average microhardness, of almost 81 ± 16.

Overall, the standard deviations in FGM specimens are similar (16, 15 and 16). The same behaviours were observed in alloy EN AC 47000, with and without Zn additions. Microhardness measured in EN AC 47000 was 95 ± 12 HV0.15 without Zn addition and 80 ± 12 HV0.15 with Zn addition.

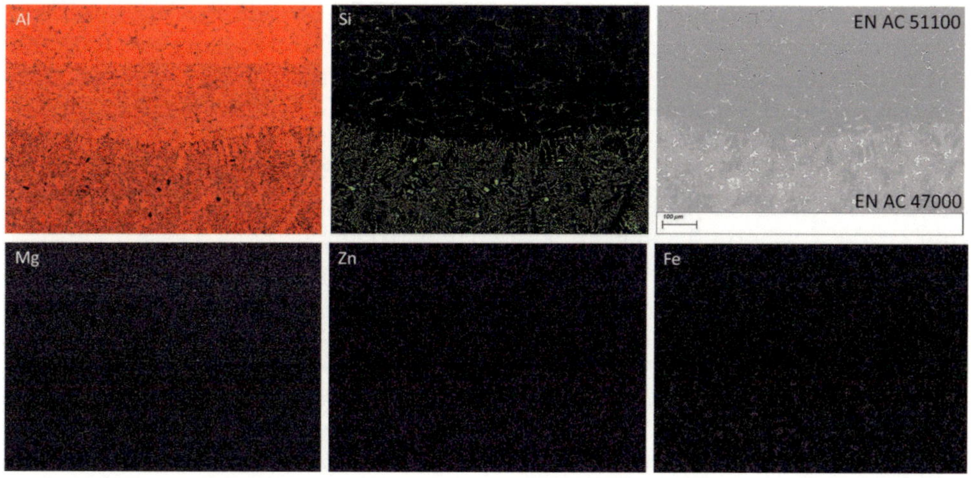

Figure 8.
EDS maps analysis in FGM (with Zn addition in both the alloys) interface.

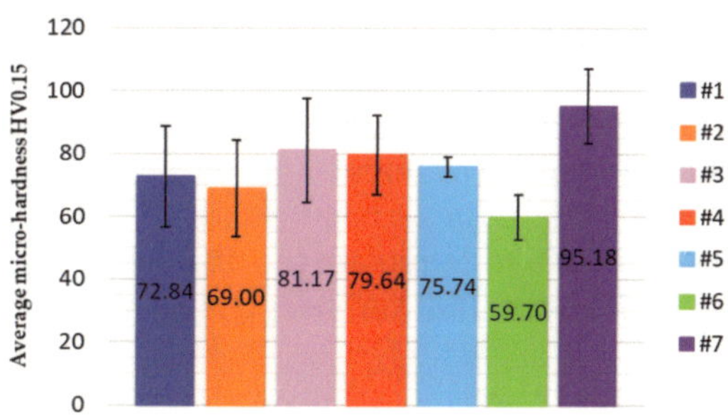

Figure 9.
Average microhardness measured in bulk alloys EN AC 47000 + Zn, EN AC 51100 + Zn, EN AC 47000, EN AC 51100 (bars 4, 5, 6, 7) and average microhardness measured for FGM, FGM with Zn addition in EN AC 47000 and FGM with Zn addition in both alloys (bars 1, 2, 3).

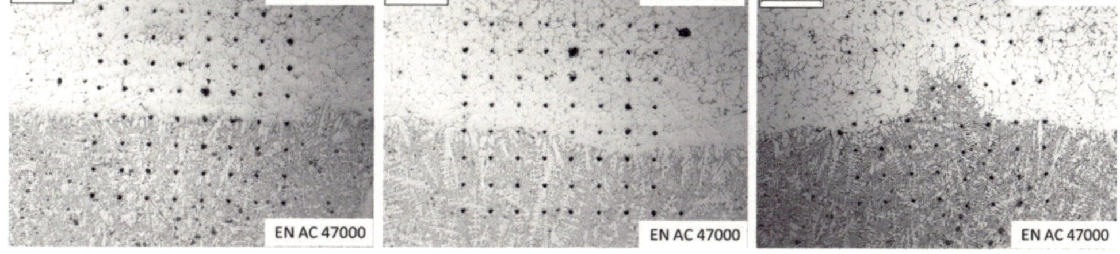

Figure 10.
Microhardness maps for FGMS. A: FGM; B: FGM with Zn addition in EN AC 47000; C: FGM with Zn addition in both alloys.

The lower microhardness in alloy EN AC 47000 after Zn addition in respect to the same alloy without Zn (bars #4 and #7, respectively) can be explained with the higher inhomogeneity detected in the alloy with Zn addition. In fact, such behaviour was not noticed in the FGMs interfaces with Zn, where microhardness resulted in improvement after Zn additions.

Considering the portion of EN AC 47000 in FGMs samples, the average microhardness measured was 78 ± 14, 74 ± 14 and 90 ± 20, respectively for FGM, FGM with Zn in EN AC 47000 and FGM with Zn in both compositions. These results seem to suggest that 1.5 w.t.% of Zn in EN AC 47000 does not significantly affect the alloy microhardness; on the other hand, Zn addition in both alloys causes an increase in microhardness of EN AC 47000 near the interface. This behaviour may be attributable to the higher Zn content detected near the interface between EN AC 51100 and EN AC 47000. The mixing of the compositions during the casting did not cause a depletion in Zn, because this constituent is present in both alloys.

Alloy EN AC 51100 has few alloying elements; thus, Zn addition may easily affect the average microhardness in this alloy, which increased from 60 ± 7 HV0.15 to 76 ± 3 HV0.15. Considering the portion of EN AC 51100 in FGM samples, the average microhardness measured were 71 ± 18, 67 ± 16 and 76 ± 11, respectively for simple FGM, FGM with Zn in EN AC 47000 and FGM with Zn in both compositions. These results suggest that Zn and the other alloying elements that characterize the alloy EN AC 47000 affect the microhardness of the alloy EN AC 51100. On the other hand, Zn addition only in alloy EN AC 47000 does not cause a further increase in EN AC 51100 microhardness, despite the composition mixing during the casting. Only the Zn addition in both alloys leads to an increase in microhardness.

The slight depletion of Mg and Zn at the interface of FGMs, detected in SEM-EDS maps in **Figures 4**, **6** and **8**, may explain the difference in microhardness measured respectively in the bulk alloy and near the FGM interfaces.

3.3.2 Tensile tests and fracture surfaces

Mechanical tensile tests were performed on FGMs specimens as well as on single alloy specimens.

Table 3 shows the tensile properties measured. FGMs are characterized by an average UTS (ultimate tensile strength) of almost 164 MPa, similar to FGMs with Zn addition. After adding Zn only in alloy EN AC 47000, the average UTS decreased to almost 151 MPa, while the elongation to fracture A% increased. In both cases, Zn addition caused a decrease of $R_{p0,2}$. In effect, in [20], the authors noticed that after adding 1.5–2% Zn, the mechanical UTS of Al-Si alloys decreases. Alloy EN AC 47000 affect the mechanical properties of all FGMs, regardless of the presence or absence of Zn. The alloy EN AC 47000 with Zn was characterized by almost 170 MPa of UTS, while the elastic module resulted very similar to the module of FGM with the addition of Zn only in EN AC 47000. Similarly, alloy EN AC 51100 with Zn has a UTS of about 148 MPa, as the UTS of FGM with the addition of Zn only in EN AC 47000. Zn addition in alloy EN AC 47000 seems not to affect UTS, resulting in slightly higher in the alloy without Zn; the same behaviour was previously noticed for the microhardness values.

EN AC 47000 has the higher average UTS, followed by EN AC 47000 with Zn and FGM without Zn. Not considering the average data, the higher UTS was detected for FGM specimen without Zn (184 MPa) while the lower was detected for alloy EN AC 51100 (122 MPa). As attended, alloy EN AC 51100 has shown the lowest mechanical properties and significant elongation to fracture A%.

In FGM specimens, it seems that $R_{p0,2}$ decreases with the addition of Zn. This behaviour goes against the solid solution hardening expected. In fact, Zn has a

$R_{p0,2}$ [MPa]	UTS [MPa]	A% [%]
FGM without Zn		
123 ± 20	164 ± 24	1.3 ± 0.6
FGM with Zn in EN AC 47000		
83 ± 0.6	151 ± 31	4 ± 3
FGM with Zn in both alloys		
99 ± 5	161 ± 20	2.8 ± 1
EN AC 47000 with Zn		
108 ± 3	171 ± 3	2 ± 0.02
EN AC 51100 with Zn		
111 ± 2	148 ± 27	2 ± 1
EN AC 51100		
73 ± 0.3	136 ± 21	6 ± 3.5
EN AC 47000		
108 ± 0.4	176 ± 9	2.5 ± 0.7

Table 3.
Average tensile tests results.

high solubility in the aluminium matrix. On the other hand, the $R_{p0,2}$ decreasing was only detected in FGMs specimens, while in alloy EN AC 47000 $R_{p0,2}$ remains almost constant and in alloy EN AC 51100 increases. In this sense, there are two possible explanations. First, the $R_{p0,2}$ decrease may be associated with a higher defect population in the FGM specimens that affect the yielding of the casting. The second hypothesis is that during the tensile test dislocations cannot pass through the Zn-straightened matrix and the primary Si particles, inducing concentrations of efforts that cause crack nucleation sites. The lower UTS after Zn addition may be similarly explained considering the additional presence of a certain defect population that affect the positive effect of Zn.

Particularly, the high standard deviation in the UTS of FGM without Zn addition was due to the presence of two specimens that fractured near casting defects. Such defects were primarily shrinkage porosities along the junction and in the proximity of the junction near the EN AC 51100. These porosities were caused by an incorrect elapsing time between the casting of the alloy EN AC 51100 and the subsequent pouring of EN AC 47000, causing air entrainment and mixing in the junction. Another important consideration must be done; notwithstanding FGM without Zn fractured preferentially in alloy EN AC 51100, the mechanical properties of the FGM without Zn still remain higher than the properties of the single alloy EN AC 51100 without Zn. This behaviour can be attributed to two reasons. The first one is that the presence of the alloy EN AC 47000, which is able to assure a high strength resistance, positively affects the FGM resistance, while the second is due to a good quality of the junction realized that provides a favourable stress distribution between the alloys.

Surface fractures were observed through SEM. In general, FGM specimens' rupture has occurred on the EN AC 51100 side; in one specimen, it occurs at the interface while, in one sample, the rupture was detected in alloy EN AC 47000 close to a defect. **Figure 11a** and **b** show fracture surfaces of two FGM without Zn. Mainly, **Figure 11a** shows the surface fracture of the FGM sample with the higher UTS, 188 Mpa. This specimen is characterized by a brittle fracture that occurred on alloy EN AC 47000. The fracture reveals a vast number of dendrites. **Figure 11b** displays an

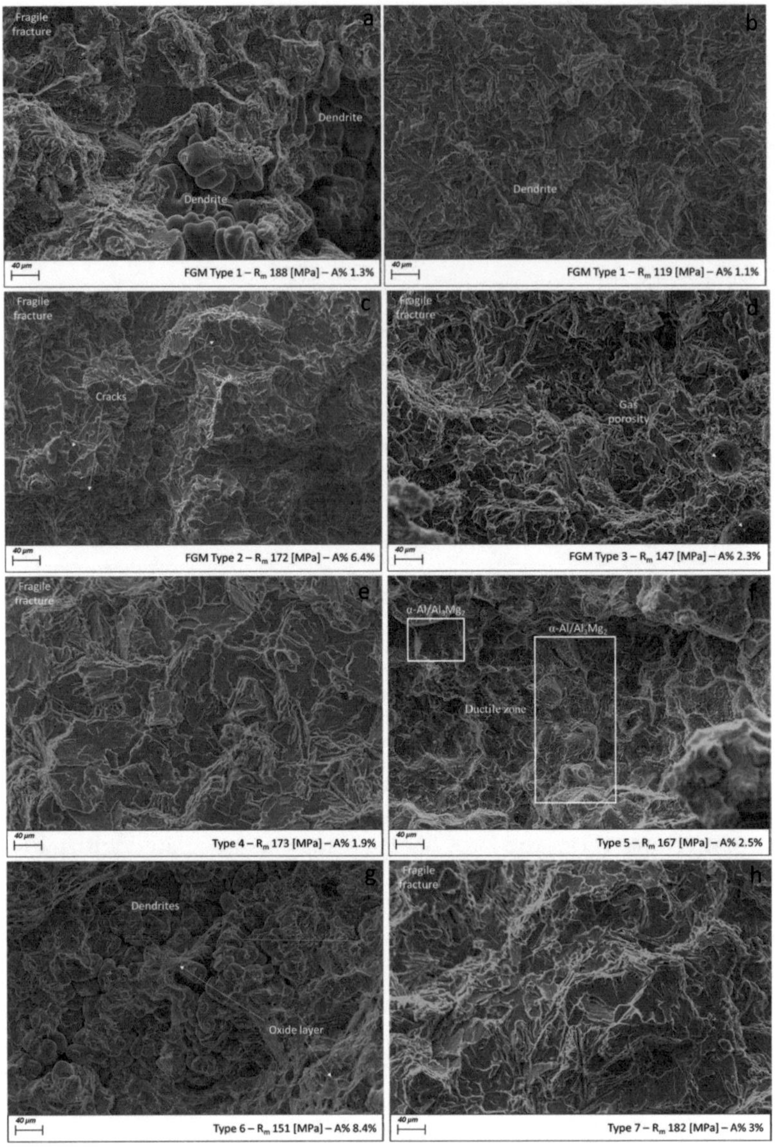

Figure 11.
Surface fractures for a few selected specimens. a, b: FGMs without Zn addition. c: FGM with Zn addition in EN AC 47000. d: FGM with Zn in both alloys. e,h: Alloys EN AC 47000 with and without Zn. f, g: Alloy EN AC 51100 with and without Zn addition.

FGM without Zn fractured along the interface between the alloys. A brittle fracture and dendrites characterize the fracture; this behaviour is symptomatic of a weak interface bonding between the two alloys. Overall, FGMs without Zn specimens are broken into the less resistant alloy, apart from specimens affected by casting issues.

Figure 11c and **d** show the fracture surfaces for both FGMs with Zn in EN AC 47000 and FGM with Zn in both compositions. Particularly, FGM with Zn only in EN AC 47000 fractured at the interface with pretty high elongation. On the other hand, FGM with Zn in both the alloys fractured with very low mechanical strength at the interface (147 MPa).

Figure 11e and **h** display the EN AC 47000, with and without Zn addition. Both fractures are brittle, and the elongation to fracture was approximately 2%. Cleavage plains were observed.

Alloy EN AC 51100 shows the highest elongation to fracture, of about 8%, while the elongation drops to about 2% after Zn addition. Despite the good mechanical properties, the elongation to fracture, in **Figure 11g**, is clearly visible extended dendrites. In **Figure 11f**, are visible the eutectic phases and a small amount of dimples.

Overall, intermetallic rod-like shapes or plate shapes cause brittle fractures for the concentration of efforts, as noticeable in **Figure 12**. Defects detected in surface fractures are mainly dendrites, but in FGM with Zn in both alloys were also noticed gas porosities. From the fracture analysis, it seems a certain grade of microporosity into the casting was realised, especially into alloy EN AC 51100. This behaviour was not previously highlighted in SEM analysis.

Near the rod-like intermetallics, are noticed cleavage plains. The very thin acicular microstructure of the eutectic regions α-Al/Al_3Mg_2 seems does not affect the rupture mode, resulting in an excellent elongation to fracture (**Figures 11f** and **12**).

3.3.3 Three-point bending test

Table 4 reports the results of bending tests for each specimen. As maximum values regard, FGM without Zn has the highest ultimate force F_{max} (394 Mpa). FGMs' deformation at rupture appears most influenced by alloy EN AC 47000 except for one sample, where the elongation resulted in about 10%.

FGM with Zn addition only in EN AC 47000 has shown the lower deformation at fracture and the lower Fmax; particularly, in sample #2, the $R_{p0,2}$ was so low that was not recorded. In FGM with Zn only in EN AC 47000, deformation at rupture is similar to deformation at rupture of alloy EN AC 51100, while in FGM with Zn in both alloys, it is similar to the deformation of alloy EN AC 47000.

Alloy EN AC 51100, as attended, has the highest deformation at rupture.

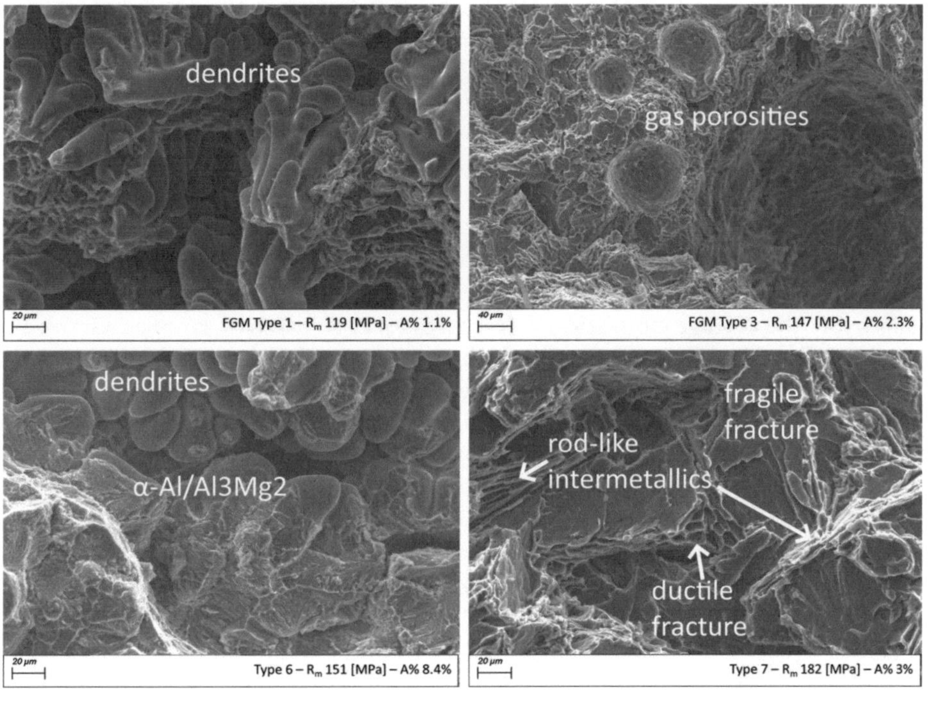

Figure 12.
Fracture details. FGM type 1 is the FGM without Zn; FGM type 3 indicates the FGM with Zn in both compositions; types 6 and 7 are alloys EN AC 51100 and EN AC 47000 without Zn.

Rp0,2 [MPa]	Fmax [MPa]	dL%
FGM without Zn		
181 ± 24	269 ± 65	5.6 ± 3
FGM with Zn in EN AC 47000		
191 ± 13	267 ± 45	4. ± 0.8
FGM with Zn in both alloys		
198*	151 ± 129 *	2.5 ± 0.2
EN AC 47000 with Zn		
183 ± 12	289 ± 24	2 ± 0.02
EN AC 51100 with Zn		
229 ± 16	306 ± 38	4 ± 0.9
EN AC 51100		
131 ± 8	318 ± 24	12.3 ± 0.9
EN AC 47000		
182 ± 11	347 ± 27	3.9 ± 0.7

Table 4.
*Three-point bending tests average results. * a brittle fracture affects both Rp0,2 (absence of yielding) and high standard deviation.*

As the three-point bending test regard, results are comparable with tensile tests. In fact, for example, FGMs in tensile conditions present similar UTS as the Fmax measured in FGM without Zn and with Zn in EN AC 47000. FGMs with Zn in both alloys resulted not comparable because of defects inside castings. Alloy EN AC 47000 has shown similar UTS with and without Zn, while in the bending test, the absence of Zn causes an increase in the specimen deformability, thanks to the lower solid solution straightening. Finally, EN AC 51100 bending behaviour is similar with or without Zn in terms of Fmax, despite the deformability decrease with Zn, as noticed for tensile test results.

3.3.4 Mechanical property connections

Average mechanical properties were compared to each other, and results are shown in graphs of **Figure 13**. **Figure 13a** shows a graph between the average ultimate tensile strength and the average elongation to fractures for each batch of samples. Graph a shows that alloy EN AC 47000 (type 4 with Zn and type 7 without Zn) has the highest R_m but low elongation to fracture. Conversely, alloy EN AC 51100 (type 5 with Zn and type 6 without Zn) have the lowest R_m, while A% changes as a function of Zn addition. Without Zn, EN AC 51100 castings reach the maximum average elongation.

As FGMs regard, Zn addition does not influence the R_m that remains almost constant, while dL% increases. The increase in dL% may be caused to the presence of nanoparticle intermetallic compounds Mg-Zn that positively affect the fracture mode, thanks to their rounded shapes [21]. Despite that, the intermetallic phases Mg-Zn are nanometric and, thus, are not easily observable at SEM. These intermetallic phases nucleate during the solidification; as the rate of Zn increases, the rate of Mg dissolved in α-Al decreases to form Mg-Zn intermetallic phases.

When comparing R_m to average microhardness values, as in **Figure 13c**, it becomes evident that EN AC 51100 castings have shown the higher R_m and higher microhardness, followed by FGM with Zn in both alloys and alloy EN AC 47000

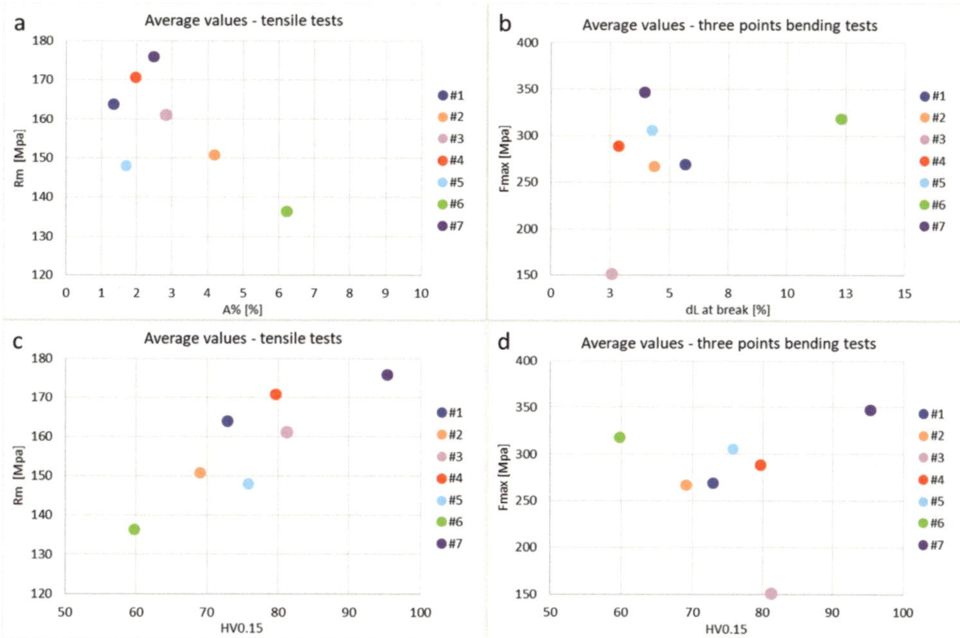

Figure 13.
Average values from **Tables 3** *and* **4**. *a: Average values Rm and A% obtained after tensile tests. b: Average values Fmax and dL% obtained after bending tests. c: Average values HV0.15 and Rm obtained after tensile tests and microhardness measurements. d: Average values HV0.15 and Fmax obtained after bending tests and microhardness measurements. FGM, FGM with Zn addition in EN AC 47000 and FGM with Zn addition in both alloys indicated as #1, #2, #3, while EN AC 47000 + Zn, EN AC 51100 + Zn, EN AC 47000, EN AC 51100 indicated as #4, #5, #6, #7.*

with Zn addition. On the other hand, as expected, EN AC 51100 castings present both the lowest R_m and microhardness. FGM with Zn only in EN AC 47000 presents a relatively low average R_m and microhardness; this was caused by a defective specimen that affects the batch's average values.

Figure 13b shows a graph between the maximum strength and the average deformation at ruptures dL% for each batch of samples. In both tensile and bending tests, EN AC 51100 has higher elongation/deformation at rupture (green points in **Figure 13a** and **b**), while EN AC 47000 specimens have the higher ultimate tensile strength/maximum force (violet points). FGM with Zn in both the alloys shows the lowest deformation at rupture. **Figure 13d** highlights the relation between the microhardness and the maximum force of bending tests; EN AC 47000 castings have the higher Fmax and higher microhardness, followed by alloy EN AC 47000 with Zn addition.

4. Conclusions

In this work, new kinds of functionally graded materials realized by controlling the mould filling were produced. The alloy Al-Si EN AC 47000 was adopted along with the Al-Mg composition EN AC 51100, obtaining castings with good-quality and low retained defects. The Zn addition was taken into account to increase the mechanical properties of the as-cast FGM, especially by nucleating the Mg-Zn intermetallic phases in composition EN AC 51100. Functionally graded materials were therefore produced in three variants—the first type was the simple casting of the two compositions; the second type involved the addition of commercially pure Zn into the Al-Si alloy, while the third type involved the Zn addition into both the alloys composing the FGM. In order to better understand the FGM properties and compare the results, samples made in purely alloys Al-Si and Al-Mg were cast, with and without Zn addition, using the same production process of FGMs.

All the castings produced were mechanically tested, and their microstructures were observed at the scanning electron microscope SEM-EDS.

FGMs, realized by alloys Al-Si and Al-Mg, presented good bonding, which the mechanical testings have highlighted. Moreover, SEM observations of the FGMs interfaces highlight, in general, the absence of extended defects.

In alloy EN AC 51100 (**Figure 2**), intermetallic phases such as Al-Fe-Mn-Si phases, the eutectic α-Al/Al_3Mg_2 phase and Mg_2Si phase were found. In alloy EN AC 47000 (**Figure 3**), intermetallic Fe-Cu-Mn phase, Q phase, polygonal eutectic silicon and Chinese script α-AlFeSiCuMg were detected.

FGM interface is shown in **Figure 4**; intermetallic phases are coherent with those observed in **Figures 2** and **3**.

The silicon diffusion into Mg-based alloy was clearly noticeable, especially in the eutectic regions α-Al/Al_3Mg_2. Moreover, Mg was detected in both the alloys with a slight depletion at the interface, despite the single alloy EN AC 47000 was not containing Mg (see **Figure 7**). In the FGM with Zn addition in both alloys, either Zn adn Mg depletion at the interface were observed. The slight depletion of Mg and Zn at the interface of FGMs may explain the difference in microhardness measured respectively in the bulk alloy and near the FGM interfaces—in bulk alloys, microhardness appears higher with respect to microhardness near the interface.

Overall, the tensile strength and the maximum bending force seems to decrease with the addition of Zn. During the tensile test, probably, dislocations cannot pass through the Zn-straightened matrix, straightening intermetallic phases and primary Si particles, inducing concentrations of efforts that cause crack nucleation sites. The lower UTS after Zn addition may be explained considering the additional presence of a certain defect population that affect the mechanical properties.

Author details

Elisa Fracchia[1*] and Mario Rosso[2]

1 Department of Applied Science and Technology, Polytechnic of Turin, Alessandria, Italy

2 INSTM c/o Department of Applied Science and Technology, Polytechnic of Turin, Alessandria, Italy

*Address all correspondence to: elisa.fracchia@polito.it

IntechOpen

© 2021 The Author(s). Licensee IntechOpen. This chapter is distributed under the terms of the Creative Commons Attribution License (http://creativecommons.org/licenses/by/3.0), which permits unrestricted use, distribution, and reproduction in any medium, provided the original work is properly cited.

References

[1] Li Y, Liu J, Zhang Q, Huang W. Casting defects and microstructure distribution characteristics of aluminum alloy cylinder head with complex structure. Materials Today Communications. 2021;**27**:102416. DOI: 10.1016/j.mtcomm.2021.102416

[2] Saleh B, Jiang J, Fathi R, Alhababi T, Xu Q, Wang L, et al. 30 Years of functionally graded materials: An overview of manufacturing methods, applications and future challenges. Composites Part B: Engineering. 2020;**201**:108376. DOI: 10.1016/j.compositesb.2020.108376

[3] Khan T, Zhang N, Akram A. State of the art review of functionally graded materials. In: Proceedings of the IEEE International Conference on Computing, Mathematics and Engineering Technologies (iCoMET 2019); 30-31 January 2019. Sukkur, Pakistan: IEEE; 2019. pp. 1-9. DOI: 10.1109/ICOMET.2019.8673489

[4] Sarathchandra DT, Kanmani Subbu S, Venkaiah N. Functionally graded materials and processing techniques: An art of review. Materials Today: Proceedings. 2018;**5**;10;1:21328-21334. DOI: 10.1016/j.matpr.2018.06.536

[5] El-Galy IM, Saleh BI, Ahmed MH. Functionally graded materials classifications and development trends from industrial point of view. SN Applied Sciences. 2019;**1**:1378. DOI: 10.1007/s42452-019-1413-4

[6] Ghanavati R, Naffakh-Moosavy H. Additive manufacturing of functionally graded metallic materials: A review of experimental and numerical studies. Journal of Materials Research and Technology. 2021;**13**:1628-1664. DOI: 10.1016/j.jmrt.2021.05.022

[7] Matuła I, Dercz G, Barczyk J. Titanium/Zirconium functionally graded materials with porosity gradients for potential biomedical applications. Materials Science and Technology. 2020;**36**(9):972-977. DOI: 10.1080/02670836.2019.1593603

[8] Vijaya Kumar P, Jebakani D, Velmurugan C, Senthilkumar V. Effect of SiC on mechanical and microstructural characteristics of Al based functionally graded material. SILICON. 2021. DOI: 10.1007/s12633-020-00933-0

[9] Marzuki M, Mazni Ismail N, Ihsan Abdul Latiff M. Preparation and microstructural characterization of five-layered aluminium-aluminium oxide functionally graded material. In: Proceedings of the 2nd International Conference on Innovative Technology, Engineering and Sciences (iCITES 2020); 22-23 December 2020. Vol. 1092. Pekan Pahang, Malaysia: IOP Conference Series: Materials Science and Engineering; 2021. p. 012031. DOI: 10.1088/1757-899X/1092/1/012031

[10] Surya MS, Prasanthi G. Effect of silicon carbide weight percentage and number of layers on microstructural and mechanical properties of Al7075/SiC functionally graded material. SILICON. 2021. DOI: 10.1007/s12633-020-00865-9

[11] Mallick A, Gangi Setti S, Sahu RK. Centrifugally cast A356/SiC functionally graded composite: Fabrication and mechanical property assessment. Materials Today: Proceedings. 2021;**47**:11. DOI: 10.1016/j.matpr.2021.07.155

[12] Elkotb HH, Mostafa R, Samad AAA, Enab TA. Manufacturing and characterization of functionally graded material automotive piston using centrifugal casting technique. Solid State Phenomena. 2021;**3**(318):13-24. DOI: 10.4028/www.scientific.net/SSP.318.13

[13] Pradeep AD, Rameshkumar T. Review on centrifugal casting of functionally graded materials. Materials Today: Proceedings. 2021;**45**:729-734. DOI: 10.1016/j.matpr.2020.02.764

[14] Karun AS, Sanil H, Rajan TPD, Pillai UTS, Pai BC. Characteristics of functionally graded bimetallic aluminium alloys by sequential casting technique. Materials Science Forum. 2015;**830-831**:383-386. DOI: 10.4028/www.scientific.net/MSF.830-831.383

[15] Kieback B, Neubrand A, Riedel H. Processing techniques for functionally graded materials. Materials Science and Engineering A. 2003;**362**(1-2):81-105. DOI: 10.1016/S0921-5093(03)00578-1

[16] Król M, Tański T, Snopiński P, Tomiczek B. Structure and properties of aluminium–magnesium casting alloys after heat treatment. Journal of Thermal Analysis and Calorimetry. 2017;**127**:299-308. DOI: 10.1007/s10973-016-5845-4

[17] Fracchia E, Lombardo S, Rosso M. Case study of a functionally graded aluminum part. Applied Sciences. 2018;**8**:7. DOI: 10.3390/app8071113

[18] Fracchia E, Gobber FS, Rosso M, Actis Grande M, Bidulská J, Bidulský R. Junction characterization in a functionally graded aluminum part. Materials (Basel). 2019;12;21:3475. DOI: 10.3390/ma12213475

[19] Ebhota WS, Jen T-C. Intermetallics formation and their effect on mechanical properties of Al-Si-X alloys. In: Aliofkhazraei M, editor. Handbook of Intermetallic Compounds - Formation and Applications. Rijeka: IntechOpen; 2017. DOI: 10.5772/intechopen.73188

[20] Němec M, Gärtnerová V, Klementová M, Jäger A. Analysis of intermetallic particles in Mg-12 wt.%Zn binary alloy using transmission electron microscopy. Materials Characterization. 2015;**106**:428-436. DOI: 10.1016/j.matchar.2015.05.038

[21] Alemdag Y, Beder M. Effects of zinc content on strength and wear performance of Al–12Si–3Cu based alloy. Transactions of the Nonferrous Metals Society of China. 2019;**29**(12):2463-2471. DOI: 10.1016/S1003-6326(19)65154-X

Chapter 2

New-Age Al-Cu-Mn-Zr (ACMZ) Alloy for High Temperature-High Strength Applications: A Review

Samarendra Roy and Shibayan Roy

Abstract

One of the prime challenges with age hardened Al-Cu alloys is the strength degradation at high temperatures (above ~250°C) due to the coarsening of strengthening θ′ precipitates and associated metastable θ′ → stable θ phase transformation. A recent discovery suggests that micro-alloying with Manganese (Mn) and Zirconium (Zr) can synergistically restrict θ′ precipitate coarsening, thereby rendering an excellent high temperature stability for Al-Cu-Mn-Zr (ACMZ) alloys. The θ′ precipitates are stabilized primarily from the reduction of interfacial energy by preferential solute segregation (Mn & Zr) at θ′ precipitate/α-Al matrix interfaces. The Al-Cu-Mn-Zr alloys thereby exhibit excellent high temperature hardness and tensile properties (yield and ultimate tensile strength) in addition to superior fatigue life and creep resistance. This newly developed Al-Cu-Mn-Zr alloys also showed excellent hot tearing resistance compared to the conventional cast Al-Cu alloys so much so that it meets the industrial standards as well. These alloys also have promising manufacturing possibility by additive route. Overall, Al-Cu-Mn-Zr alloys offer great potential for the automotive industry because of their unprecedented high temperature performance which should enable engineers to build light weight passenger vehicles leading to a safer and greener environment.

Keywords: Al-Cu-Mn-Zr alloy, precipitate strengthening, high temperature stability, solute segregation, mechanical property, additive manufacturing

1. Introduction

Aluminum alloys have been one of the most prominent structural material system for many years now; this is also reflected in their global usage, they come only next to steel [1, 2]. Owing to their high specific strength, resistance to stress corrosion cracking, excellent fatigue resistance, workability and cost effectiveness [3], Aluminum alloys are one of the primary material choice for aerospace and automotive industries [2, 4]. In addition, Aluminum based composites were developed over the years to mitigate some of the limitations of Aluminum alloys and further facilitates their use for various engineering applications [5]. Apart from structural applications, Aluminum and its alloys are also employed for the electronics and electrical industries in abundance for their suitable combination of functional properties [6].

Pure Aluminum is characterized by low yield strength which is improved by many folds from different strengthening strategies e.g. by adding different alloying elements, thereby making the alloys suitable for structural applications. Depending upon the major alloying element/s, Aluminum alloys are classified in two important categories; some of the alloys can be strengthened by heat treatment (age hardening) while others by mechanical deformation (non-age hardenable) [7, 8]. Al-Cu alloys fall in the first category; they are strengthened by in situ precipitates introduced through appropriate thermal treatment (aging). In age hardening, these precipitates hinder the dislocation motion and increase the hardness or strength of the alloy [9].

Despite the beneficial attributes, age-hardening Aluminum alloys, especially Al-Cu alloys suffer from limited high temperature capability. At temperature above 200°C, the metastable strengthening precipitates (e.g. θ') undergo rapid growth and coarsening and even transform to stable incoherent precipitates (metastable $\theta' \rightarrow$ stable θ precipitate) which are inefficient to restrict dislocation motion. This leads to a concurrent steep decrease in their high temperature load bearing capabilities. Due to such degradation, Al-Cu alloys are restricted for elevated temperature applications in spite of their light weight and high specific strength at room temperature.

In recent times, a new class of Al-Cu alloys is developed by suitable micro-alloying with Mn and Zr which possesses excellent stability for strengthening θ' precipitates at and above 300°C. The Al-Cu-Mn-Zr, termed as ACMZ alloys, provide significant improvement in most of the elevated temperature mechanical properties including hardness, tensile strength, creep and fatigue resistance etc. The present chapter provides a detailed account of the development of Al-Cu-Mn-Zr alloys while highlighting the limitations of existing Al-Cu alloys in the first place. It alongside discusses about the underlying mechanisms responsible for their excellent high temperature stability and subsequently on various properties. The chapter finally access the possibility of industrial adaptation of this newly developed alloy system and expected industrial impacts in long run.

2. Precipitate formation and evolution in Al-Cu system

During age hardening of Al-Cu alloys, it is first heat-treated at temperatures where single-phase α-Al solid solution is formed; this process is known as solution treatment [7]. Afterwards, the alloy is rapidly quenched to room temperature which causes freezing of solute Cu atoms within the α-Al matrix, thus forming a super saturated solid solution (SSSS). The solute atoms afterwards can diffuse even at room temperature through this super-saturated α-Al matrix and form various Al-Cu precipitates; this process is known as natural aging. In this regard, **Figure 1a** represents the binary Al-Cu phase diagram along with the solvus lines for various metastable precipitate phases. However, in common engineering practice, the solutionized alloy is heat treated at certain elevated temperature, usually within the two-phase region to produce various metastable precipitates depending on the heat treatment time and temperature (artificial aging). At sufficiently low aging temperature or during natural aging, Cu solute cluster first form from the quenched-in vacancies within the supersaturated α-Al matrix. These solute clusters then arrange in a single layer of Cu atoms known as GP -I zones (Guinier Preston zone) along a plane parallel to the (001) plane of the α-Al matrix.

When the aging process continues above the GP zone solvus line (**Figure 1a**), θ'' metastable precipitates (also known as GP-II zones) form by diffusion of Cu atoms from the solute cluster zones (GP -I zone). θ'' are characterized by the chemical formula Al_3Cu and ordered tetragonal structure ($a = b = 4.04$Å$, c = 7.68$Å) having two layers of Cu atoms separated by three layers of Al atoms (**Figure 1b**) [10]. It is completely

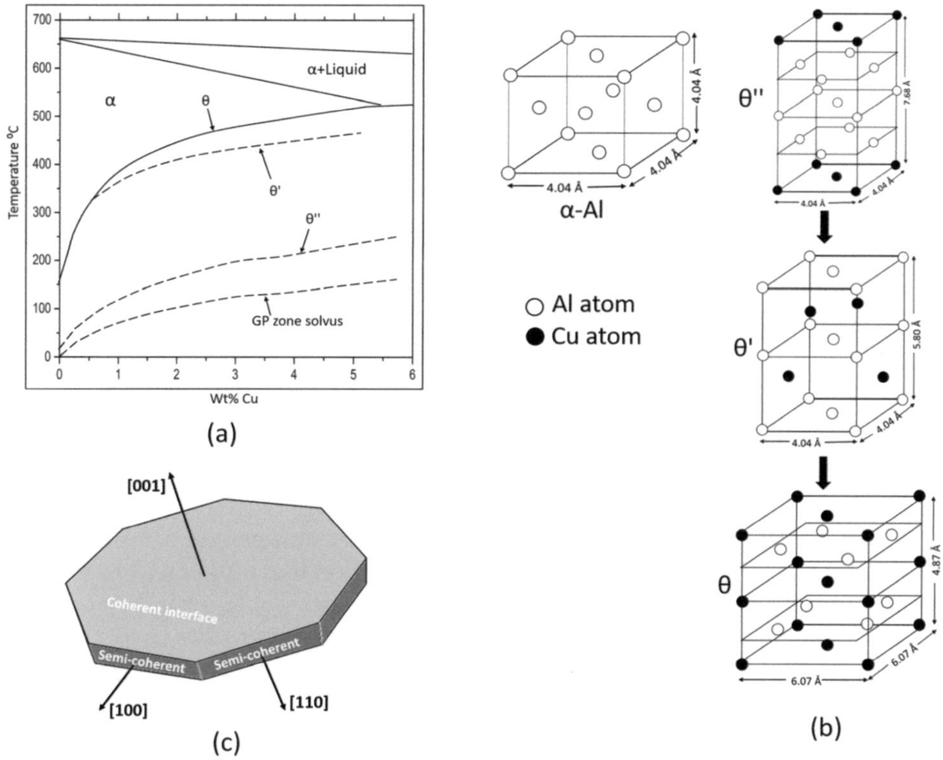

Figure 1.
Schematics showing (a) section of Al-Cu binary phase diagram, (b) crystal structure of parent α-Al matrix and various metastable (θ″ and θ′) and stable (θ) precipitates in Al-Cu alloy system, (c) morphology and different interfaces for metastable θ′ precipitate.

coherent with α-Al matrix from all sides although misfit strain develops along $(100)_{Al}$ and $(010)_{Al}$ planes. Metastable θ′ precipitates next form from θ″ precipitates on further aging. θ′ has a stoichiometric formula of Al_2Cu with body centred tetragonal structure ($a = b = 4.04Å, c = 5.80Å$, space group I4/mmm) and carries 3 layers of Al atom and 2 layers of Cu atoms in an ordered arrangement (**Figure 1b**) [10, 11].

In the θ″→θ′ transformation, the later precipitate starts nucleating at the pre-existing θ″ precipitates via continued diffusion of Cu atom and continues to grow until the entire θ″ precipitate transforms into θ′. In this transformation, θ′ holds the same orientation relationship with α-Al matrix as does θ″ precipitate i.e. $(001)_{\theta'} \| (001)_{Al}$ and $[100]_{\theta'} \| [100]_{Al}$ [12]. It grows as a plate within the α-Al matrix parallel to the $(010)_{Al}$ and remains coherent along this plane. However, along the $(100)_{Al}$ and $(010)_{Al}$ planes, θ′ becomes incoherent or complex semi-coherent with α-Al matrix. The semi-coherent side of the plate shaped θ′ precipitate is not completely circular; rather it forms as octagon with facets along $[100]_{\theta'}$ and $[110]_{\theta'}$ directions (**Figure 1c**). The $\{110\}_{\theta'}$ interface edges further act as a solute gateway from where Cu atoms diffuse and coarsen θ′ precipitates on prolonged thermal exposure. The atomic arrangement as well as ledge dislocations on $\{110\}_{\theta'}$ semi-coherent interfaces further assist in the accommodation of Cu atoms [13].

The equilibrium precipitate in the Al-Cu system is tetragonal θ ($a = b = 6.07Å, c = 4.87Å$, space group I4/mcm) having chemical formula Al_2Cu (**Figure 1b**) [10]. It is incoherent with the α-Al matrix along all sides. θ forms from θ′ precipitate via mechanism similar to θ″ → θ′ transformation. The habit plane of θ also remains parallel to (100) α-Al matrix planes.

The entire precipitation process during aging of Al-Cu alloy therefore can be summed up as: Solute clusters → GP zones → GP II zones (θ″) → θ′ → θ.

3. Strengthening mechanisms in Al-Cu alloy system

The operating strengthening mechanism/s in Al-Cu alloy system differs as a function of precipitate type, mainly with their size and coherency with α-Al matrix. When the precipitates are small and coherent with α-Al matrix (e.g. GP -I or θ″), it is energetically easy for the dislocations to shear through them. This usually occurs via one or a combination of mechanisms like chemical strengthening, stacking-fault strengthening, modulus strengthening, order strengthening, coherency strengthening etc. [14]. Apart from chemical strengthening, the increment in critical resolved shear stress (CRSS) varies with the size of the precipitates as $\sim r^{1/2}$ where r is the precipitate radius. In case of chemical strengthening, CRSS increment is inversely proportional to r.

When the precipitates (usually θ′) are large in size and possess semi-coherent or incoherent interfaces with α-Al matrix, the dislocation line bulges within the inter-precipitate region rather than shearing through the precipitate until they meet and move forward while leaving a dislocation loop behind (**Figure 2a**) [16]. This process is known as Orowan looping. The increase in CRSS (for spherical precipitates) due to Orowan looping is given by:

$$\Delta\tau = \left(\frac{3}{2\pi}\right)^{\frac{1}{2}} \cdot \frac{Gb}{r} \cdot f^{\frac{1}{2}} \tag{1}$$

where, $\Delta\tau$ is the increase in CRSS due to Orowan strengthening, G is the shear modulus of the α-Al matrix and f is the precipitate volume fraction. The CRSS increment thus varies with $1/r$. Furthermore, for plate shaped θ′ precipitate forming on $\{100\}_{Al}$ habit plane having diameter D and thickness t, CRSS increment, $\Delta\tau$ due to Orowan looping is given by [17]:

$$\Delta\tau = \left(\frac{Gb}{2\pi\sqrt{1-\nu}}\right)\left(\frac{1}{0.931\sqrt{\frac{0.306\pi DT}{f}} - \frac{\pi D}{t} - 1.061T}\right)\left(ln\,\frac{1.225t}{r_0}\right) \tag{2}$$

where, ν is the Poisson's ratio of α-Al matrix and r_0 is the radius of the dislocation core.

The CRSS increment with respect to the precipitate radius due to either particle shearing or Orowan looping is schematically represented in **Figure 2b** [12]. It seems that a critical radius exists for the strengthening precipitate below which particle shearing is preferred. When the precipitate grows beyond this critical radius, the dislocations prefer to bow around the precipitate rather than shearing it.

Considering finely dispersed coherent θ″ precipitates within α-Al matrix, they cause strength increment by one or a combination of mechanisms mentioned above. As these coherent precipitates grow, a corresponding increase occurs in the alloy strength since CRSS increment is proportional to $r^{1/2}$ for most of the strengthening mechanisms. At the critical radius, the strength is highest, also signifying for θ″ → θ′ transformation. Afterwards, dislocation bowing around large and semi-coherent θ′ precipitates is the main strengthening process. On further coarsening, the number density continuously decreases for θ′ precipitates so that the alloy

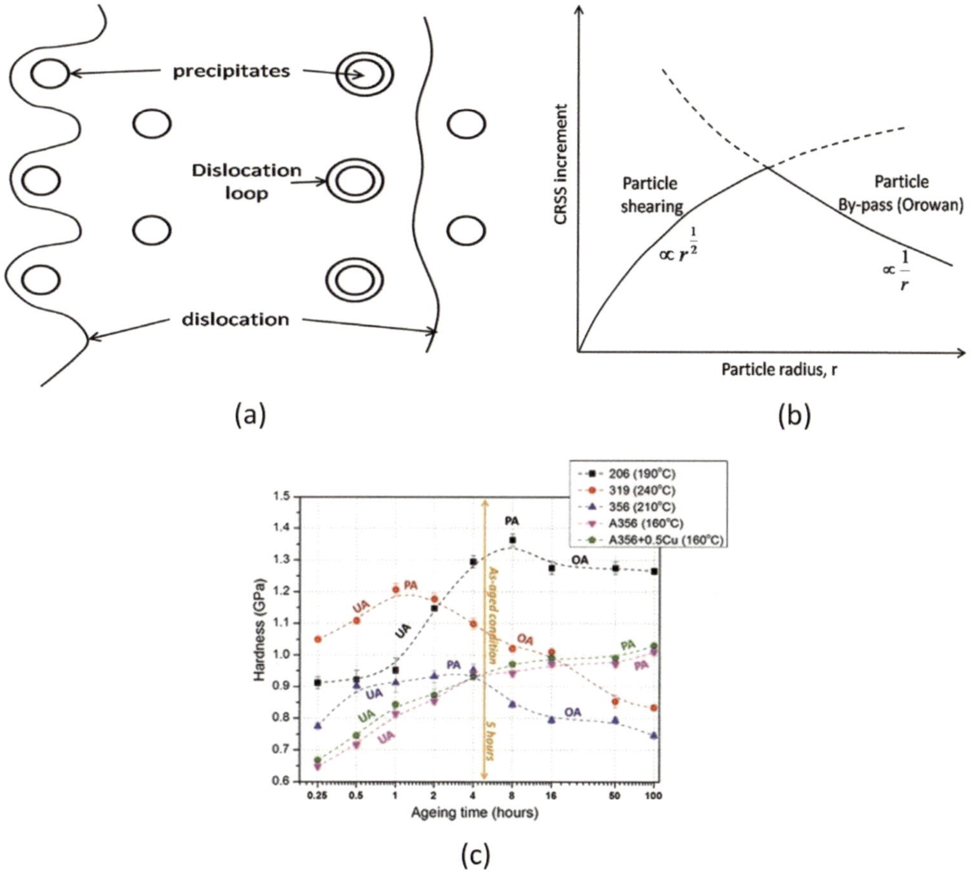

Figure 2.
Schematic showing (a) the mechanism of Orowan looping, (b) CRSS increment as a function of precipitate radius for particle shearing and Orowan looping mechanisms, and (c) isothermal aging curves (hardness vs. aging time) for various commercial Al alloys e.g. Al-Cu (206), Al-Si-Cu (319) and Al-Si (356, A356, A356 + 0.5Cu) [15].

strength correspondingly decreases since CRSS increment is proportional to *1/r* in Orowan looping.

4. Age hardening and aging curve

At the beginning of aging treatment, fine and uniform precipitation of GP -I and/or θ'' occurs within the super saturated α-Al matrix. These precipitates hinder the dislocation motion and increases the strength (or hardness) compared to the solutionized and quenched alloy (**Figure 2c**) [7, 8]. The number density of the precipitates is less and inter-precipitate distance is correspondingly high at this condition so that the alloy strength/hardness is marginally increased; the alloy is said to be in under-aged condition. Shearing of coherent GP-I and θ'' precipitates is the primary strengthening mechanism here. As the aging time increases, GP -I zones continuously transform to θ'', thereby increasing their number density in the α-Al matrix. Some of them even transform to other metastable precipitates e.g. θ'. Together, the strength/hardness increases further with continued aging due to shearing of coherent and semi-coherent precipitates [18].

With increase in aging time, aspect ratio and number density of θ′ precipitates continuously increase because of which precipitate shearing becomes more and more difficult; the strength/hardness of the alloy also keeps increasing gradually at this stage [19]. At a critical precipitate size, the strength (or hardness) of the alloy reaches a peak denoting the peak-aged condition. The θ′ precipitate size, aspect ratio, inter-precipitate distance etc. are now optimum for maximum hindrance towards dislocation motion due to shearing through the precipitates.

The θ′ precipitates continuously coarsen with increasing aging time and even start transforming to stable θ precipitate under prolonged aging [12]. The energy required to shear θ′ precipitates becomes quite high at this stage so that rather than shearing, dislocations prefer to bow around them (Orowan looping) [8]. The strength/hardness of the alloy thereafter decreases with increasing aging time leading to over-aged condition when metastable θ′ continue transforming to stable θ precipitate. **Figure 2c** represents typical aging curves (hardness vs aging time) for several commercial age-hardening Al alloys e.g. Al-Cu (206), Al-Si-Cu (319) and Al-Si (A356, A356 + 0.5Cu). In these aging curves, hardness of the alloys initially increase (under-aged condition), reaches the maximum at peak-aged condition and decreases again with subsequent aging leading to the over-aged condition [12].

Furthermore, the ductility (usually expressed in terms of elongation to fracture) of age-hardening Aluminum alloys also varies with aging time in accordance to the size, morphology and coherency of the strengthening precipitates [20, 21]. When the precipitates are small and coherent and their number density is low, dislocations can move past easily through them leading to maximum contribution from strain hardening that delays fracture. Correspondingly, under-aged alloy shows maximum ductility. On the other hand, semi-coherent and large strengthening precipitates at the peak aged condition renders maximum hindrance to the dislocation motion due to Orowan looping resulting in significant pile up at the precipitate sites. The ductility for the peak-aged alloy is also correspondingly minimum. With over-aging, some ductility is restored since the coarse, incoherent precipitates are generally not suitable to hinder dislocation motion and little pile up results around them. Overall, the ductility vs. aging time variation follows a reverse trend to the strength/hardness vs. aging time curves.

5. Challenge for high temperature stability of Al-Cu alloy: precipitate coarsening

One of the major hindrances for widespread use of Al-Cu alloys, especially in the automobile and aerospace sectors, is their poor high temperature stability associated with rapid decrease in load bearing capacity above ~250°C [22]. This strength degradation is because of the rapid coarsening of θ′ precipitates from thickening along the broad facets, thereby resulting in a drastic decrease in their aspect ratio. At high temperature, the rate of diffusion for solute Cu atoms increases so that they segregate at the coherent and semi-coherent interfaces of θ′ precipitates [23]. The broad facet (i.e. the coherent interface) thickens by ledge formation while the semi-coherent interfaces having high interfacial energy grow from accommodation of Cu atoms along $\{100\}_{\theta'}$, $\{010\}_{\theta'}$, $\{110\}_{\theta'}$ interfaces [24]. The thermodynamic driving force for coarsening of metastable θ′ precipitates is the reduction of interfacial energy [13]. It should be noted however that the increase in diameter of θ′ precipitate by the growth of semi-coherent interface does not essentially reduce the interfacial energy; rather the thickening of the broad facet i.e. growth of the coherent interface is responsible for the decrease in interfacial energy during coarsening.

At long thermal exposure, θ′ precipitates ultimately transform to stable (equilibrium) θ precipitates, which are completely incoherent with the parent α-Al matrix. Further thermal treatment leads to the growth of larger θ precipitates in expense of the smaller ones. This corresponds to an increase in the inter-precipitate distance so that coarse θ precipitates are no longer effective in restricting the motion of matrix dislocations; hence, the strength of the alloys decreases drastically [9]. It therefore appears imperative to stabilize strengthening metastable θ′ precipitates against coarsening for successful high temperature application (beyond ~250°C) of Al-Cu alloys.

6. Attempts towards the development of high temperature Aluminum alloys

Over the years, alloying pure Aluminum with various elements showed good promises for elevated temperature applications. Such elements included rare earths (e.g. Erbium, Ytterbium, Scandium etc.) as well as Zirconium, Silicon etc. [25–27]. For these alloys, formation of coherent precipitates with cubic $L1_2$ crystal structure and reduced interfacial energy is the key for their high temperature stability against precipitate coarsening upon thermal exposure [26]. The other unique exploration is the formation of core-shell structure for the strengthening precipitates which also provides excellent coarsening resistance through the minimization of interfacial energy. For example, addition of 0.06 at% Zr or 0.03 at% Er in Aluminum individually form ordered $L1_2$ Al_3Zr or Al_3Er precipitates and show moderate coarsening resistance at elevated temperature [25]. Simultaneous addition of Zr and Er in similar quantity however, leads to $Al_3(Er, Zr)$ precipitates with unique core shell structure, which made them coarsening resistant up to 400°C for 750 hours. The difference in diffusivity between Zr and Er was held responsible for formation of such core-shell structure; while Er having higher diffusivity forms the primary precipitate with Aluminum, slower diffusing Zr segregates later at the interfaces of these primary precipitates resulting in the core-shell structure. Similarly, addition of Sc to Al-Zr-Sc-Er alloy (concentrations of both Sc and Er are 0.06 at%) leads to a dual shell layer of Zr and Sc according to their respective diffusivity in Al matrix over the Al_3Er core precipitate [26].

Furthermore, excellent creep resistance was observed for Al-0.1 at% Zr and Al-0.1 at% Zr-0.1 at% Ti alloy systems at 300°C, 350°C and 400°C, which is attributed to the high temperature stability of Al_3Zr precipitates [28]. Out of the two alloys, ternary Al-Zr-Ti alloy showed comparatively lower creep resistance than binary Al-Zr alloy due to the lower lattice parameter mismatch between $Al_3(Zr_{1-x}Ti_x)$ core-shell precipitates with the parent α-Al matrix. The addition of Yb similarly resulted in excellent thermal stability for Al-0.9 at% Zr- 1.73 at% Yb alloys having $Al_3(Zr,Yb)$ precipitates up to 400–425°C [29].

For Al-Si system, Al-Si-Cu-Mg alloys are traditionally used for making high temperature pistons for automobile engines [30, 31]. These alloys show satisfactory microstructural stability as well as fatigue resistance at high temperatures which are essential requirements for automotive applications. A viable route for further improving their high temperature performance is by addition of transition metals that forms thermally stable intermetallic precipitates. For example, controlled Zr addition (up to 0.11 wt%) increases the ultimate tensile strength (UTS) of Al-Si-Zr piston alloys by 3.8% at 350°C due to the alteration in the morphology of strengthening ZrAlSi precipitates from flake to block shape [32, 33]. However, increase in Zr content up to 0.46 wt% resulted in a decrease in UTS by 5%. Similarly, A356 alloy (Al-7Si-0.4 Mg) modified with 0.25 wt% Er and nominal amount of Zr (0 to 0.6 wt%)

showed improved high temperature mechanical properties [34]. With increase in Zr content up to 0.59 wt%, both hardness and tensile strength increases at room and elevated temperatures due to the formation of $Al_3(Er,Zr)$ precipitates.

Hypoeutectic Al-7 wt%Si-1wt%Cu-0.5 wt% Mg alloys also shows excellent retention of hardness and tensile strength up to 240–260°C when micro-alloyed with 0.15 wt% Zr, 0.28 wt% V and 0.18 wt% Ti [35]. Further exposure to 475°C upto 128 hours led to additional improvement in hardness which can be attributed to the accelerated precipitation of $Al_3(Zr,V,Ti)$ and Q′ precipitates. Similarly, addition of minor Ti (0.22 wt%), Zr (0.39 and 0.19 wt%) and Ni (0.46–0.21 wt%) to commercial 354 alloy showed improvement in tensile properties up to 300°C compared to the base alloy [36]. In both cases, micro-alloying elements synergistically result in unique and complex precipitate formation which improved the high temperature stability of the corresponding alloys. For hypereutectic Al-Si alloys, Ni addition up to 1–4 wt% to Al-12 wt%Si-0.9 wt% Cu-0.8 wt% Mg alloy resulted in retention of room temperature mechanical properties, including creep resistance up to 250°C due to the formation of thermally stable Al_3Ni precipitates [37]. In addition to primary and eutectic Si, incorporation of 1 wt% ZnO nanoparticles (particle size ~40 nm) also enhance the high temperature tensile strength and elongation for Al-20 wt% Si alloys [38].

In case of age hardening Al-Cu system, several attempts were made in the past to increase their high temperature stability by adopting various strategies. Lin et al. [39] studied the effect of Ni addition (0.5–1.5 wt%) on the elevated temperature mechanical properties of squeeze cast Al-Cu-Mn-Fe alloys. At 300°C, the amount of thermally stable precipitates (e.g. Al_9FeNi, Al_3CuNi and $Al_{20}Cu_2Mn_3$) increases with increasing Ni content which enhances the elevated temperature mechanical properties of the base alloy. Addition of La in Al-Cu alloy similarly results in the formation of $Al_{11}La_3$ precipitates leading to a better high temperature mechanical properties with 0.3 wt% La being the optimized concentration [40]. The addition of 1.6–2.0 wt % Li also shows excellent mechanical properties for AA2099 (Al-Cu-Li) alloys at high temperature, primarily due to the enhanced thermal stability of T_1 (Al_2CuLi) precipitates compared to other possible strengthening precipitates like θ′ and S (Al_2CuMg) [41]. At higher temperature, T_1 precipitates coarsen instead of dissolving unlike θ′ or S. In addition, AA2219 alloy possesses improved high temperature performance when micro-alloyed with 0.8 wt% Sc, 0.45 wt% Mg and 0.2 wt% Zr from grain refinement and simultaneous precipitation of Al_3Sc, Al_3Zr and Ω precipitates along with other common strengthening precipitates like θ′ and θ″ [42].

Another viable strategy of increasing the thermal stability of Al-Cu alloys is by micro-alloying with various secondary elements for stabilization of strengthening metastable θ′ precipitates. For example, 0.18 at% Sc addition forms thermally stable Al_3Sc precipitate wherein Sc atoms tend to segregate at the α-Al matrix/θ′ precipitate interfaces [43]. Such co-stabilization of two different precipitates renders excellent high temperature property for Al-Cu alloys with Sc addition [44]. Micro alloying with Zr provides similar effects as Sc; Zr forms stable Al_3Zr precipitates having $L1_2$ structure which can act as a heterogeneous nucleation site for θ″ precipitates, thus generating a finer scale microstructure for Al-Cu-Zr alloys [45]. The resultant alloy shows excellent thermal stability up to 250–300°C. In Al-Cu-Mg alloys, 0.09–0.13 wt% Mg is reported to accelerate the formation of θ′ precipitates and the resultant alloy exhibits excellent thermal stability at 300°C for 1000 hours [46]. Similarly, addition of 0.45 wt% Ce provides heterogeneous nucleation sites for precipitation of Ω precipitates and enhances the thermal stability of Al-5.3% Cu-0.8% Mg-0.6% Ag (wt.%) alloys by retarding the diffusion of Cu [47].

Overall, there have been numerous efforts in the past to design high temperature Aluminum alloys from different binary systems (Al-Cu, Al-Si etc.), primarily by micro-alloying with various elements. However, most of these attempts showed certain shortcomings. The working temperature of the resultant ternary or quternary alloys could not be increased above 300°C under prolonged exposure. Also, use of exotic elements like rare earth additions hindered their industrial acceptance and commercial viability. Hence, the demand of cost-effective Aluminum alloys for high temperature applications has only increased over the years without much of a success.

7. Development of Al-Cu-Mn-Zr alloy

As mentioned before, age hardening Al-Cu alloys faces significant precipitate coarsening, which restricts their use for high temperature applications [48]. Numerous attempts have made over the years to increase the operational temperature for Al-Cu alloys; the most successful approach was by trace addition of various elements like Sc and Zr [49]. Micro-alloying improves the high temperature stability in two distinct ways:

a. Micro alloying elements can provide heterogeneous nucleation sites for primary strengthening precipitates (e.g. θ'' and θ') so that relatively finer precipitates with narrow size distribution is obtained in the room temperature microstructure [50]. For instance, when Al-Cu alloys are micro-alloyed with Sn, the semi-coherent interface of metastable θ' precipitate nucleates from Sn particle [51].

b. Micro alloying elements can segregate at the high energy mobile semi-coherent precipitate/matrix interfaces and at times, at the less-mobile coherent interfaces [52]. Such segregation eventually results in the reduction of energy for these interfaces, making them difficult to grow. The semi-coherent interfaces is usually more effected by such segregation [53, 54]. In recent times, stabilization of semi-coherent and coherent interfaces of strengthening θ' precipitates yield a new series of high temperature Al-Cu alloys with unprecedented thermal stability up to 350°C and beyond [48]. These alloys contain micro-alloying addition of Manganese and Zirconium and designated as Al-Cu-Mn-Zr or ACMZ alloys.

The classical approach for developing any new alloy system relies on the age-old trial and error method which has serious drawbacks, primarily considering the resource constrains and added cost from industrial standpoints [55, 56]. A more state of the art strategy of alloy designing is by using integrated computational materials engineering (ICME) approach [57, 58]; a successful example of this is realized in the development of Al-Cu-Mn-Zr alloys. The key components of ICME approach for the development of this alloy system are: (a) thermodynamic and kinetic approximations for stability of precipitates against growth controlling mechanism/s, (b) appropriate modeling for assessment of thermo-physical and thermo-mechanical properties from existing phases, (c) simulation and model/s to predict defect formation during casting processes, (d) models for prediction of microstructure during casting and other thermo-mechanical processing operations, (e) models for property prediction from microstructure and defect structure evolution, and (f) models for manufacturing of components at in-service conditions [59, 60].

These above mentioned steps were followed in the development of Al-Cu-Mn-Zr alloys which was primarily aimed to replace traditional Al-Si and Al-Si-Cu alloys in automotive applications (e.g. cylinder heads in passenger vehicle engines) [59]. Firstly, thermo-physical and thermo-mechanical properties for casting process simulation were obtained from thermodynamic databases. The simulation of casting process was conducted to estimate the casting defects and as-cast microstructure. Afterwards, thermodynamic models were employed for optimization of heat treatment cycle in terms of desired precipitation sequence, precipitate growth etc. This helped to estimate the spatial variation of thermo-physical properties over the component scale as well. Finally, component level properties e.g. residual stress, fatigue performance etc. were evaluated in order to estimate the in-service performance of the alloys considering the above mentioned parameters [61].

8. High temperature stability of Al-Cu-Mn-Zr alloy

The primary mechanism for high temperature stability of Al-Cu-Mn-Zr alloys is related to the segregation of micro-alloying solute atoms (Mn and Zr) at θ′ precipitate/α-Al matrix interfaces [48, 52]. Although it seems fairly straight forward in the first go, the interface stabilization process exhibits extreme intricacies throughout the entire precipitation sequence. As denoted earlier, the primary strengthening precipitate in Al-Cu system is θ′, which has a plate shaped morphology where the broad facets are coherent with parent α-Al matrix (**Figure 1c**) [12]. The rim of the precipitates, on the other hand, are semi-coherent and have a higher interfacial energy compared to their coherent counterparts which makes them highly mobile and prone to coarsening [48].

On exposure to high temperature for an extended duration, θ′ precipitates coarsen due to the enhanced diffusion of solute Cu atoms [8]. Segregation of Mn and Zr atoms at the θ′ precipitate/α-Al matrix interfaces prohibits Cu diffusion and further coarsening at elevated temperature [48]. The main driving force behind the solute segregation is the reduction of precipitate/matrix interfacial energy, especially for the semi-coherent interfaces. In addition, several other mechanisms like solute drag, ledge poisoning etc. also helps in the stabilization of θ′ precipitates [52, 62]. These mechanisms are explained individually below.

8.1 Segregation of micro-alloying elements

In the earliest report on Al-Cu-Mn-Zr alloys, Shyam et al. [48] compared two cast Al-Cu-Mn-Zr alloys having nominal compositions Al-5Cu-1.5Ni-0.2Mn-0.17Zr and Al-6.4Cu-0.19Mn-0.13Zr (in wt%) with conventional Al-Cu (206) and Al-Si-Cu (319) i.e. with non- Al-Cu-Mn-Zr alloys containing negligible concentration of Zr (**Figure 3**). At room temperature, base Al-Cu and Al-Si-Cu alloys exhibit superior mechanical response (higher yield strength and ductility) than Al-Cu-Mn-Zr alloys. However, the trend completely reverses after treating the alloys at higher temperature (300°C) for 200 hours; Al-Cu-Mn-Zr alloys now represent superordinate mechanical response than either Al-Cu or Al-Si-Cu alloys. Microstructural examinations reveal that θ′ precipitates significantly coarsen and transform to thermodynamically stable θ precipitates for base Al-Cu or Al-Si-Cu (i.e. non Al-Cu-Mn-Zr) alloys because of which their mechanical properties degraded after thermal treatment. On the other hand, θ′ precipitates retain their morphology and aspect ratio on high temperature heat treatment in case of Al-Cu-Mn-Zr alloys.

Bahl et al. [52] further studied the aging kinetics and thermal stability of Al-Cu-Mn-Zr alloys and showed that they retained their room temperature mechanical properties even after exposure at 300°C for 5000 hours. During such prolonged thermal treatment, θ′ precipitates suffer limited decrease in number density up to 200 hours. No further significant decrease was observed, and the peak-aged microstructure remains fairly stable up to 5000 hours.

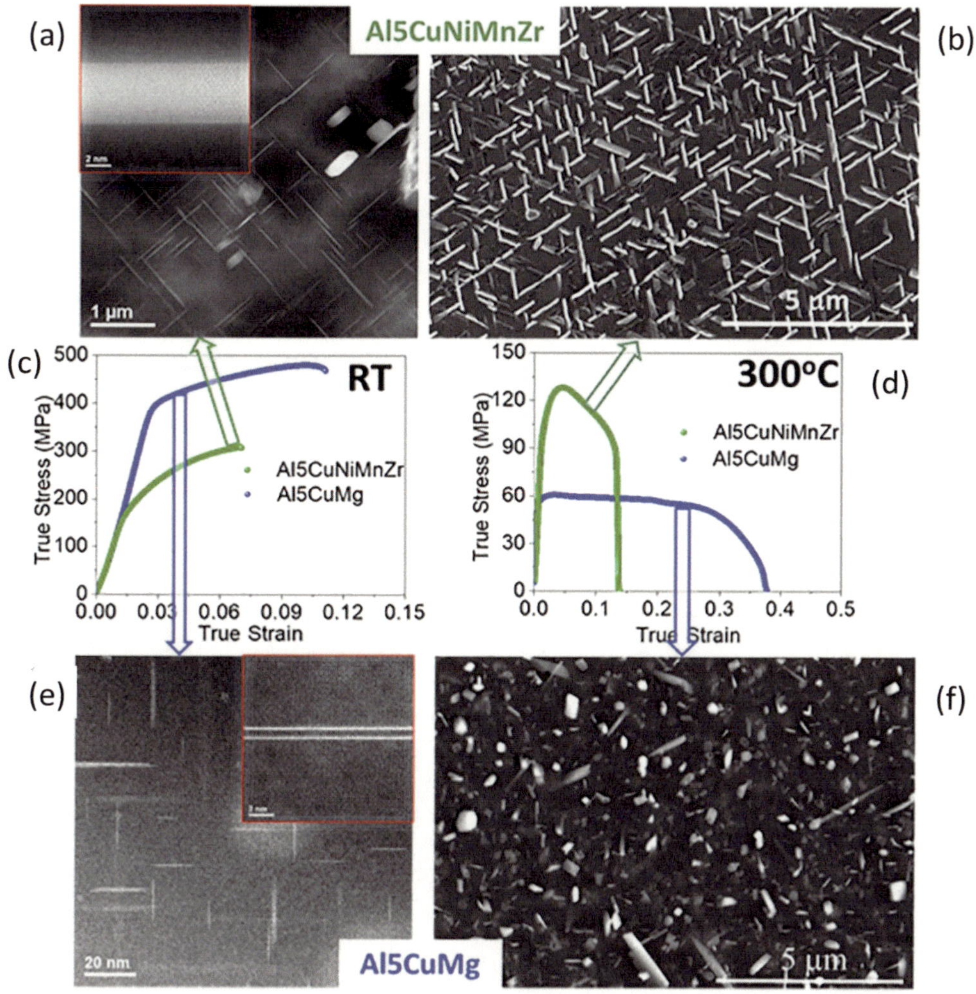

Figure 3.
(a) and (b) showing the microstructures of Al-Cu-Mn-Zr alloy in peak aged condition and post 300°C thermal exposure for 200 hours, respectively; (c) and (d) represents true-stress-true strain curves for these alloys from tensile tests carried out at room temperature and 300°C, respectively; (e) and (f) showing the microstructures for conventional Al-Si-Cu alloy under similar conditions [48].

The stability of θ′ precipitates in Al-Cu-Mn-Zr alloy was examined using atom probe tomography (APT) characterization which are shown in **Figures 4a** and **b** (side view and top view, respectively) [48, 62]. **Figure 4c** represents corresponding composition profile which indicates segregation of Cu, Mn, Zr and Si at the coherent and semi-coherent θ′ precipitate/α-Al matrix interfaces for Al-Cu-Mn-Zr alloys after prolonged (200 hours) thermal exposure at 300°C. As it appears, Mn tends to segregate both at the coherent and semi-coherent interfaces of θ′ precipitates with the segregation tendency being larger at the later interfaces. Zr, on the other hand, segregates more on the corner rim of the coherent/semi-coherent interfaces, although certain extent of Zr segregation also occurs on these interfaces. Silicon have similar segregation profile as Mn; it can in fact influences the solute segregation at these interfaces to a much greater extent as discussed later [63].

In order to understand the individual and synergistic effect of Mn and Zr micro-alloying on the thermal stability of Al-Cu-Mn-Zr alloy, a consolidated study was carried out by Poplawsky et al. [62] on several model alloys (e.g. Al-Cu-Mn, ACM and Al-Cu-Zr, ACZ etc.) in addition to the base Al-Cu-Mn-Zr alloy. The Al-Cu-Mn

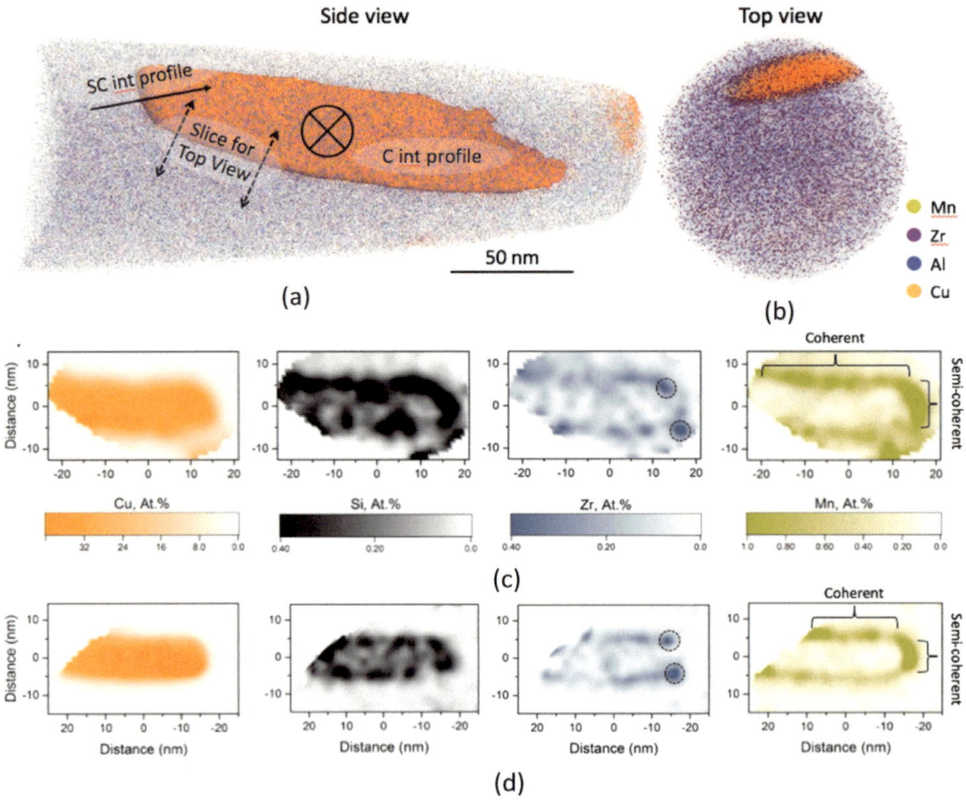

Figure 4.
(a) and (b) APT compositional maps (iso-concentration surfaces) representing side and top views, respectively of APT needle for Al-Cu-Mn-Ni-Zr alloy pre-conditioned at 300°C for 200 hours, (c) and (d) showing 2D contour plots for Cu, Si, Zr and Mn atoms on the cross-sectional planes of θ′ precipitate along <110> direction from Al-5Cu-Ni-Mn-Zr and Al-7Cu-Mn-Zr alloys, respectively after pre-conditioning at 300°C for 200 hours [48].

alloys retain their room temperature mechanical strength after exposure at 300°C for 200 hours whereas Al-Cu-Zr alloys could sustain their stability only up to 200°C. For comparison, Al-Cu-Mn-Zr alloys are stable up to 350°C. The trend in Mn segregation for Al-Cu-Mn-Zr alloy in this case is similar to that observed earlier by Shyam et al. [48] up to 300°C. Larger Mn segregation occurs at semi-coherent interfaces while minor segregation at the coherent interfaces.

After thermal exposure at 350°C, Mn segregation at semi-coherent interfaces becomes insignificant, which aggravates the mechanical degradation of Al-Cu-Mn alloys at this temperature range [62]. At 350°C, Mn tends to diffuse within the bulk of θ′ precipitates, thereby causing even lesser segregation at the semi-coherent interfaces. Zr, on the other hand, retains their segregation profile at the coherent interfaces up to 200°C for Al-Cu-Zr alloys. The Zr segregation profile is also similar in nature to that observed previously for Al-Cu-Mn-Zr alloys by Shyam et al. [48].

8.2 Diffusional perspective

Solid state diffusion is one of the key component for evolution of precipitate structure and morphology during the course of thermal exposure [64]. In the corresponding binary systems, self-diffusion coefficient of Cu is much higher than that for Mn so that diffusion of Cu atoms continue to coarsen θ′ precipitates unless interfacial segregation resists [63]. Furthermore, self-diffusion coefficient of Zr in

Al is almost 10 times lower than that for Mn in Al [65, 66]. As a result, Mn atoms diffuse much faster to the coherent and semi-coherent interfaces of θ′ precipitates than Zr atoms during the initial thermal exposure for peak-aged Al-Cu-Mn-Zr alloys. Mn segregation thereafter pins both the interfaces and restricts the flux of Cu atoms from coarsening the θ′ precipitates on further heat treatment. The slow diffusing Zr atoms, on the other hand, tend to segregate mostly at the corner of the coherent and semi-coherent interfaces at a later stage of heat treatment and contributes to their stabilization only during prolonged thermal exposure.

Due to this sequence of segregation (initial segregation of Mn followed by Zr segregation on prolonged thermal exposure), θ′ precipitates are stable only up to 300°C in Al-Cu-Mn (ACM) alloys where Mn alone is the micro-alloying element. In the absence of Zr, Mn segregation is insufficient to restrict coarsening at 350°C since they diffuse into the bulk of the precipitates rather than segregating at the interface at higher temperature or up to prolonged thermal exposure [62]. On the other hand, Zr being a slowly diffusing element, requires longer duration or higher temperature to segregate at the interface of α-Al/θ′ precipitates. However, sufficient coarsening of θ′ precipitates may have already occurred or they may even transform to the stable θ precipitate by the time Zr stabilizes the interfaces on prolonged thermal exposure. In addition, Zr segregation preferentially takes places at the coherent interfaces which has lower mobility compared to the semi-coherent ones. Hence Zr segregation alone is least efficient to stabilize θ′ precipitate and the stability of Al-Cu-Zr (ACZ) alloys is limited only up to 200°C. In case of Al-Cu-Mn-Zr (ACMZ) alloys, Mn atoms initially stabilize both coherent and semi-coherent interfaces. The slow diffusing Zr atoms thereafter segregate at the coherent interfaces and provides further stabilization of θ′ precipitates. This sequential segregation of Mn and Zr synergistically provides stability for Al-Cu-Mn-Zr alloys up to 350°C and beyond for a prolonged duration [48].

8.3 Effect of precipitate size and interparticle spacing

The diffusion aided coarsening of θ′ precipitates can be best described using the classic Lifshitz-Slyozov-Wagner (LSW) theory [67] where the rate of coarsening depends on the corresponding mass transfer mechanism (lattice diffusion, interface atomic mobility, grain boundary diffusion, pipe diffusion through dislocation cores etc.). The governing equation in LSW theory is given as [68]:

$$\bar{r}^3 - \bar{r}_0{}^3 = k.t \tag{3}$$

where, $\bar{r}$ & $\bar{r}_0$ are the mean precipitate radius at time $t = t$ and $t = 0$, respectively and k is a constant. Partial derivation of Eq. (3) with respect to t gives:

$$\frac{\partial \bar{r}}{\partial t} = \frac{k}{3\left(k.t + \bar{r}_0{}^3\right)^{\frac{2}{3}}} \tag{4}$$

Since $\frac{\partial \bar{r}}{\partial t}$ varies with the initial precipitate radius as $\sim \frac{1}{\bar{r}_0{}^2}$, the smaller the initial precipitate radius, the higher will be the rate of its coarsening. In case of peak-aged Al-Cu-Mn-Zr alloy, θ′ precipitates are considerably larger at room temperature with higher inter-precipitate distance. Their larger size helps to reduce the coarsening rate on thermal exposure since higher inter-precipitate distance ensures non-overlapping diffusion fields [48].

Furthermore, the constant k in Eq. (4) can be represented as $k = D_{Cu}\gamma_{sc}X_e$, where D_{Cu} is the diffusional coefficient of Cu in Al, γ_{sc} is the interfacial energy of semi-coherent interface of θ′ precipitate and X_e is the equilibrium solubility of Cu in Al [52]. Reduction in the interfacial energy of semi-coherent interfaces due to solute segregation therefore helps in decreasing the value of *k*. This in turn contributes to the reduction in the coarsening rate for θ′ precipitates.

As it seems, the microstructural requirement for better coarsening resistance and high temperature stability of Al-Cu-Mn-Zr alloy is quite counterintuitive. At room temperature, a fine-scale microstructure with smaller precipitates and correspondingly, smaller inter precipitate spacing is preferred for high strength [8]. However, a larger precipitate with higher inter-precipitate spacing is desired for enhanced coarsening resistance at higher temperature. Together, Al-Cu-Mn-Zr alloys present low to moderate strength at room temperature but excellent retention of that strength at elevated temperature [48].

8.4 Role of trace elements (Si and Ti)

Other than the major micro-alloying elements (Mn & Zr), trace elements (e.g. Si and Ti) present in the composition may further influence the microstructural stability of Al-Cu-Mn-Zr alloy at elevated temperature. Silicon decreases the coarsening resistance for θ′ precipitates so that Si content in Al-Cu-Mn-Zr alloys should be preferably below 0.1 wt% [48]. Si being a faster diffusing species than Zr and even than Mn, preferentially occupies atomic positions at θ′ precipitate/α-Al matrix interfaces above this critical concentration (>0.1 wt%), thereby preventing further segregation of Mn or Zr atoms at these locations. However, Si is not as efficient as Mn or Zr for stabilization of θ′ precipitates at elevated temperatures. Below the threshold concentration, presence of Si is not detrimental for Al-Cu-Mn-Zr alloys as shown by Shower et al. [63]. Si content in the range of 0.05 wt% to <0.1 wt% can even outperform the hardness of base Al-Cu-Mn-Zr alloys at elevated temperature.

After solution treatment of Al-Cu alloys, quench-in vacancies can cluster together to form edge dislocations at room temperature [12, 69]. When these vacancies are in significant density, they can even form dislocation loops rather than individual dislocations, which can further climb and form dislocation helices [70–74]. These helices accommodate far more number of vacancies with their spacings being larger than individual dislocations. When Si atoms are present in significant quantity in the binary Al-Cu alloy, they can also cluster together during aging due to high diffusivity. The dislocations, dislocation helices and Si clusters can all potentially provide heterogeneous nucleation sites for θ′ precipitates when aged above θ″ solvus.

When the Si content is kept low (<0.05 wt%), Al-Cu alloys essentially act as a binary system and θ′ precipitates mostly nucleate at the dislocation loops, thereby promoting a fine scale microstructure on aging [63]. At higher Si content (0.11 wt %-0.24 wt%), θ′ precipitates nucleate at Si clusters, which again leads to a finer microstructure. However, at the intermediate Si content (0.05–0.1 wt%), nucleation of θ′ precipitates primarily occurs at the dislocation helices. As a result, the number density of θ′ precipitates decreases such that their inter-precipitate spacings become larger as well as the critical size for θ′ → θ transformation increases compared to either high or low Si containing Al-Cu alloys. At room temperature, such coarse microstructure of Si containing Al-Cu alloys yields low hardness in peak-aged condition. However, initially larger θ′ precipitates tend to coarsen far less during elevated temperature exposure since their larger size and greater inter-particle spacings provide better resistance.

Titanium when present in trace concentration can also influence the high temperature stability of Al-Cu-Mn-Zr alloys by forming stable Al_3Ti precipitates having $L1_2$ crystal structure [75]. Titanium atoms show similar segregation profile as Zr at θ′ precipitate/α-Al matrix interfaces. Poplawsky et al. [62] in this regard observed unique $L1_2$ structured $Al_3(Zr_xTi_{1-x})$ precipitates on the α-Al matrix/ θ′ precipitate interfaces from addition of Ti to Al-Cu-Mn-Zr alloys. The (001) interfaces of this $Al_3(Zr_xTi_{1-x})$ precipitates are coherent with $(100)_{Al}$ as well as $(001)_{\theta'}$ planes, which helps to stabilize θ′ precipitates by reducing their misfit strain. Such $L1_2$ precipitate formation further restricts addition of Cu atoms to θ′ precipitates, thereby retarding their thickening along the coherent interfaces. This mechanism of coarsening resistance is known as "ledge poisoning" [63]. The semi-coherent interfaces, on the other hand, are depleted of $Al_3(Zr_xTi_{1-x})$ precipitates. This is further confirmed and explained from DFT calculations considering interfacial energetics of preferential precipitation [62, 63]. The formation of $Al_3(Zr_xTi_{1-x})$ precipitates on the coherent interface in turn creates θ′/$Al_3(Zr_xTi_{1-x})$ /α-Al structure which is energetically favorable compared to similar structures on the semi-coherent interfaces.

Furthermore, self-diffusion coefficients of Mn, Zr and Ti in Al vary in the order $D_{Mn} > D_{Zr} > D_{Ti}$ so that Mn atoms diffuse faster and segregate at the coherent and semi-coherent interface of θ′ precipitates on thermal exposure while slower diffusing Zr atoms segregates primarily at the coherent interfaces [63]. In the peak-aged condition, semi-coherent interface of θ′ precipitate is therefore populated only with faster diffusing Mn atoms. On thermal exposure at 300°C for 200 hours, Mn atoms initially segregate at both coherent and semi-coherent interfaces. Afterwards, the slower diffusing Zr atoms segregate at the coherent interfaces. During further heat treatment at 350°C for 200 hours, Ti atoms can diffuse and form $Al_3(Zr_xTi_{1-x})$ precipitates on the coherent interfaces as well as at the edges of coherent/semi-coherent interfaces; the relative proportion of these precipitates remains higher on the former location. During the high temperature heat treatment, Mn on the other hand, tend to penetrate towards the bulk of θ′ precipitates. This sequence of segregation of various micro-alloying elements is schematically depicted in **Figure 5**.

8.5 Computational studies

As discussed before, reduction in interfacial energy due to the segregation of solute atoms (Mn and Zr) at the mobile interfaces promotes thermal stabilization for metastable θ′ precipitates up to a prolonged duration [48]. In this regard, density functional theory (DFT) simulations were carried out to determine the

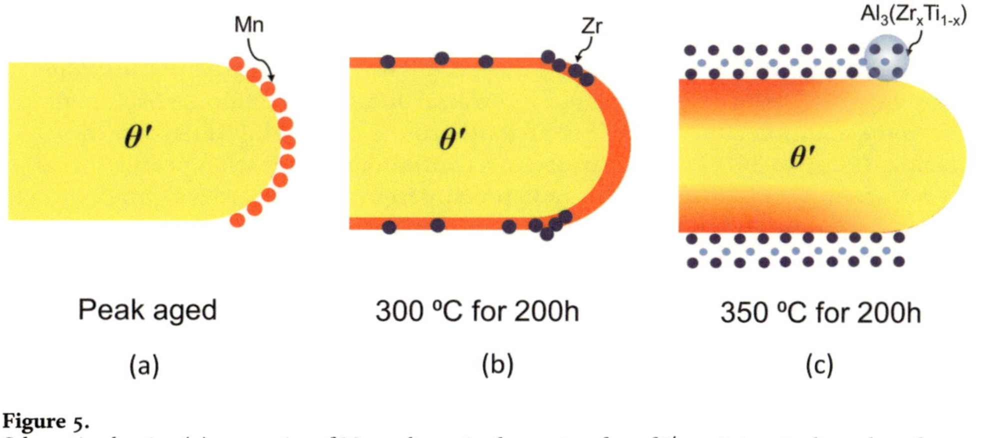

Figure 5.
Schematics showing (a) segregation of Mn at the semi-coherent interface of θ′ precipitate in the peak aged condition, (b) Zr segregation after prolonged thermal exposure at 300°C for 200 hours and (c) formation of $Al_3(Zr_xTi_{1-x})$ precipitates on the edges of coherent/semi-coherent interfaces as well as penetration of Mn through the bulk of θ′ precipitate after heat treatment at 350°C for 200 hours.

interfacial energy for various θ′ precipitate/α-Al matrix interfaces with and without solute addition (Mn & Zr). In DFT calculations, segregation energy *(ΔE_{seg})* was defined as

$$\Delta E_{seg} = \Delta E_{sol}(int) - \Delta E_{sol}(bulk) \quad (5)$$

where, $\Delta E_{sol}(int)$ and $\Delta E_{sol}(bulk)$ is the heat of solution when certain solute element situates at θ′ precipitate/α-Al matrix interface and in the bulk of the precipitate, respectively. The interfacial energy change is then defined as:

$$\Delta\gamma = \frac{\Delta E_{seg}}{2A} \quad (6)$$

where, A is the area of the interface. The calculations suggest that the interfacial energy for pure (i.e. without any segregation) coherent and $(100)_{\theta'}$ semi-coherent interfaces are 252 mJ/m^2 and 527 mJ/m^2, respectively [48]. It further establishes that the addition of Mn and Zr in Al-Cu-Mn-Zr alloys decreases the interfacial energy for both coherent and semi-coherent interfaces. It also corroborates with the experimental observations that Mn atoms preferentially segregate at the semi-coherent interfaces while Zr atoms occupies places both at coherent and semi-coherent interfaces [62].

Important to note that the coarsening of strengthening precipitates is ***NOT*** a function of interfacial energy alone, although reduction in interfacial energy certainly adds as a dominating factor for hindering the precipitate coarsening [76, 77]. Lattice misfit strain is another important driving force for coarsening of θ′ precipitates. Other kinetic factors like diffusion barrier formation and solute drag are crucial too for restricting coarsening of θ′ precipitates on thermal exposure [52, 78]. Due to the formation of diffusion barrier around θ′ precipitates, mobility of Cu atoms within α-Al matrix reduces to a large extent, which further helps to prevent coarsening or transformation of these precipitates. In addition, the presence of slow diffusing element/s in any Al alloy system can contribute to the coarsening resistance of θ′ precipitate [79]. Such slower diffusing element creates a solute drag since they need to be displaced during the growth of the interfaces, which in turn retards the coarsening of θ′ precipitate.

Introduction of a third element (Mn, Zr etc.) within the binary Al-Cu alloys can lead to one or a combination of thermodynamic and kinetic restrictions (mentioned above) to precipitate coarsening processes. Shower et al. [76] in this regard carried out a phase field modeling study to understand the synergistic effect of various mechanisms that offers precipitate coarsening resistance. The study suggests that a combination of interfacial energy reduction and solute drag due to the addition of Mn and Zr contributes to the coarsening resistance of θ′ precipitates up to 300°C. In the process, a continuous segregation profile forms for Mn atoms along the interfaces of θ′ precipitates with a larger weightage at the semi-coherent interface. Other solute atoms e.g. Zr, which introduce a positive misfit strain do not effectively interact with the mobile semi-coherent interfaces; rather they tend to segregate at the coherent interfaces. When the working temperature is raised to 400°C, resistance to precipitate coarsening requires simultaneous reduction in the mobility of Cu atoms through α-Al matrix and the interfacial energy of θ′ precipitates. This could not be achieved by micro-alloying with Mn and Zr alone, which is why the Al-Cu-Mn-Zr alloys losse their excellent thermal stability at 400°C and beyond. All the cumulative and inter-connected effects that contribute to the stabilization of θ' precipitate at elevated temperature in Al-Cu-Mn-Zr alloy are schematically shown in **Figure 6**.

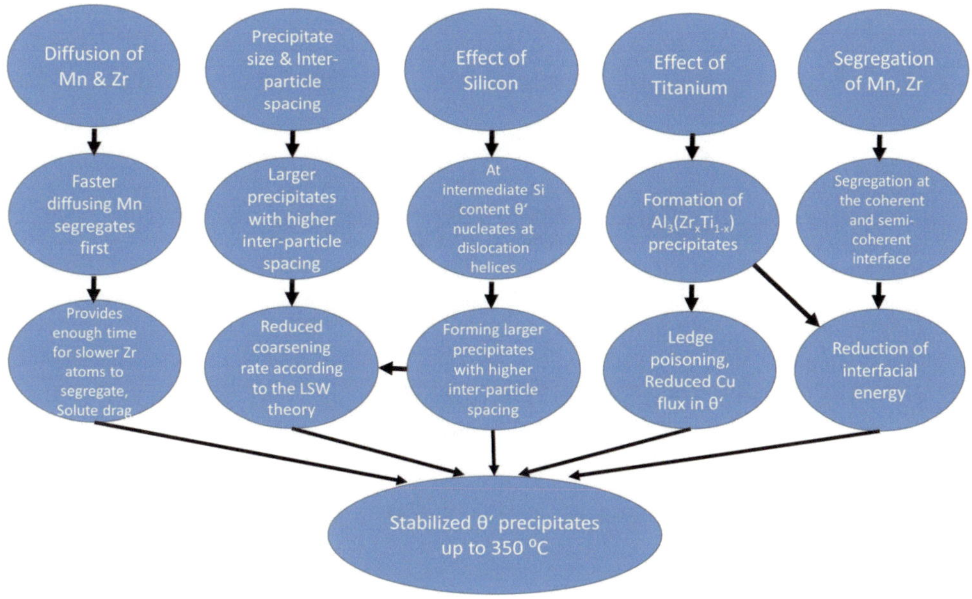

Figure 6.
Schematic flowchart showing the cumulative effect of various contributing factors and mechanisms involved in the stabilization of θ′ precipitates at elevated temperatures for Al-Cu-Mn-Zr alloys.

9. Mechanical properties of Al-Cu-Mn-Zr alloy

9.1 Hardness and tensile properties

Similar to other age hardening Al-Cu alloys, primary strengthening mechanism for peak-aged Al-Cu-Mn-Zr alloys is Orowan looping where matrix dislocations bow around the coarse θ′ precipitates [52]. Apart from this, solid solution and grain boundary strengthening (by Hall–Petch mechanism) also contributes to the overall strength of the Al-Cu-Mn-Zr alloys. Analytical calculations, however, suggest that Orowan looping plus other strengthening mechanisms together are inadequate to account for experimentally measured yield strength of Al-Cu-Mn-Zr alloys [52]. This calls for the consideration of additional strengthening mechanisms e.g. stress-free transformation strain (SFTS). The formation of θ′ precipitates in the α-Al matrix is usually associated with transformation strain fields which can interact and potentially restricts dislocation movement, thereby increasing the alloy strength further.

Figures 3c–d indicates that the room temperature tensile properties for peak aged Al-Cu-Mn-Zr alloy is inferior compared to the conventional peak-aged Al-Cu alloy. For example, the ultimate tensile strength(UTS) of Al-5Cu alloy is ~490 MPa whereas it is ~300 MPa for Al-Cu-Mn-Zr alloy at room temperature [48]. In addition, the yield strength of the later alloy is nearly half compared to the base Al-Cu alloy. However, after prolonged thermal exposure at 300°C, the trend reverses; Al-Cu-Mn-Zr alloy possess nearly twice the UTS and yield strength compared to the base Al-Cu alloy. Similarly, Al-5Cu-Mg alloy possess higher hardness than Al-Cu-Mn-Zr alloys at room temperature (**Figure 7**). With increase in pre-conditioning temperatures (heat treatment for 200 hours), non- Al-Cu-Mn-Zr alloys show drastic decrease in hardness around 200°C, while Al-Cu-Mn-Zr alloys can sustain the room temperature hardness without any significant degradation until 350°C.

Bahl. et al. [52] further showed that the hardness and yield strength of peak-aged Al-Cu-Mn-Zr alloy drop marginally during post-aging thermal exposure but remained

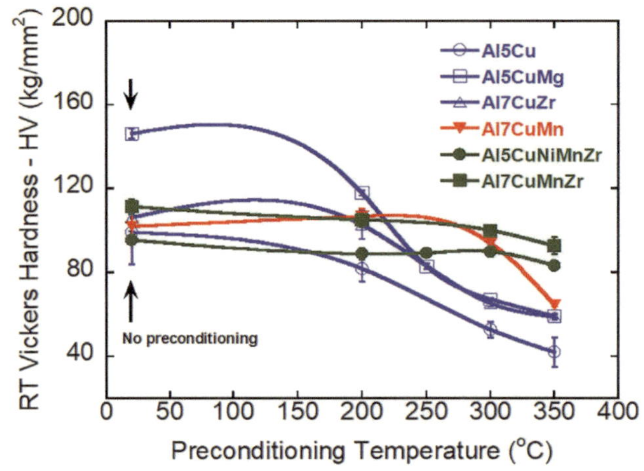

Figure 7.
Room temperature hardness for various conventional Al-Cu and Al-Cu-Mn-Zr alloys as a function of pre-conditioning temperatures [48].

almost constant during prolonged thermal treatment up to 5000 hours. This accounts for a stable microstructure with almost constant precipitate volume fraction, thickness, diameter, aspect ratio, equivalent diameter, number density and inter-precipitate spacing for Al-Cu-Mn-Zr alloys on extended thermal exposure. The ductility of Al-Cu-Mn-Zr alloys are further influenced from Cu content although it does not vary the yield strength and UTS much. For example, increasing the Cu content from 6 wt% to 9 wt% causes the fracture strain to reduce by 50% primarily due to the increased amount of brittle intermetallics at α-Al matrix grain boundaries [80].

Important to note that no comprehensive study is yet to report the results pertaining to full-scale tensile testing, especially the strain hardening response as well as the fracture characteristics of Al-Cu-Mn-Zr alloys whether at room or elevated temperatures. The earliest available work of Shyam et al. [48] showed certain true stress–strain curves for Al-Cu-Mn-Zr alloy from tensile tests carried out at room temperature and 300°C in comparison to regular Al-Cu alloy (**Figure 3**). The purpose of the tensile tests was however, to establish the superiority for the former alloy at elevated temperature. Important to note that the alloys (Al-Cu and Al-Cu-Mn-Zr alloys) were used in peak-aged condition for room temperature tensile tests and after pre-conditioning at 300°C for 200 hours for elevated temperature tests.

The true stress–strain curves from room temperature tensile tests suggest that peak-aged Al-Cu-Mn-Zr alloy possesses marginally higher strain hardening rate compared to the conventional Al-5Cu-Mg alloy at least in the initial part of the plastic regime. The hardening rates although do not vary much at the later part (below UTS) representing almost similar slopes for both alloys. During 300°C tensile tests, both Al-Cu-Mn-Zr and Al-5Cu-Mg alloys exhibit substantial strain softening, however at significantly differing rates; the rate of softening is greater for Al-Cu-Mn Zr alloy compared to the conventional Al-Cu-Mg alloy. The ductility for the former alloy is also always higher than the later alloy irrespective of the test temperature. The strain hardening response for precipitate hardened systems at room temperature is generally attributed to the isotropic hardening of α-Al matrix plus kinematic hardening due to dislocation pile up at the precipitate locations from continued Orowan looping [81–83]. The strain softening at elevated temperature can possibly be attributed to dynamic recovery of piled-up dislocations which reduces dislocation density at precipitate cites and increases ductility by delaying the final fracture.

9.2 Creep response

The excellent high temperature stability of Al-Cu-Mn-Zr alloys enhances their creep properties as well. Miligan et al. [84] conducted creep tests under different stress levels for various Al-Cu-Mn-Zr alloys with varying grain sizes at 300°C and compared their creep resistance with base Al-Cu alloy as well as Al-Sc alloy which is known for its excellent creep resistance. **Figure 8** represents the steady state creep strain rate as a function of applied stress for these alloys. At low stress level, stress exponents for Al-Cu-Mn-Zr alloys are close to unity signifying for diffusional creep being the dominant mechanism. On the other hand, dislocation creep is the mechanism for conventional Al-Cu alloy as identified from a higher stress exponent. The dislocation movement though α-Al grain interiors is difficult at low stress levels for Al-Cu-Mn-Zr alloys due to the enhanced thermal stability of θ′ precipitates; rather grain boundary diffusion dominates at high temperature making diffusional Coble creep as the rate controlling mechanism. At higher stress levels however, the controlling mechanism switches to dislocation creep even for Al-Cu-Mn-Zr alloys since the grain boundary precipitates effectively slow down the movement of vacancies. This in turn restricts grain boundary diffusion as well as grain boundary sliding.

9.3 High temperature deformation response

One of the prime motivations for the development of Al-Cu-Mn-Zr alloy is to replace conventional cast Aluminum alloys (e.g. Al-Si-Cu based 319 alloy) for making light weight components in automotive engines [48, 85]. The Al-Cu-Ni based RR350 alloy with 0.2 wt% Mn and 0.17 wt% Zr, which can be considered as a variant of Al-Cu-Mn-Zr alloys, is also used as light weight and high temperature resistant alloys for high end automobile engine applications over the years [86]. Shower et al. [87] has compared the effect of microstructural stability on the high temperature deformation response of RR350 alloy vis-à-vis 319 alloy in as-cast

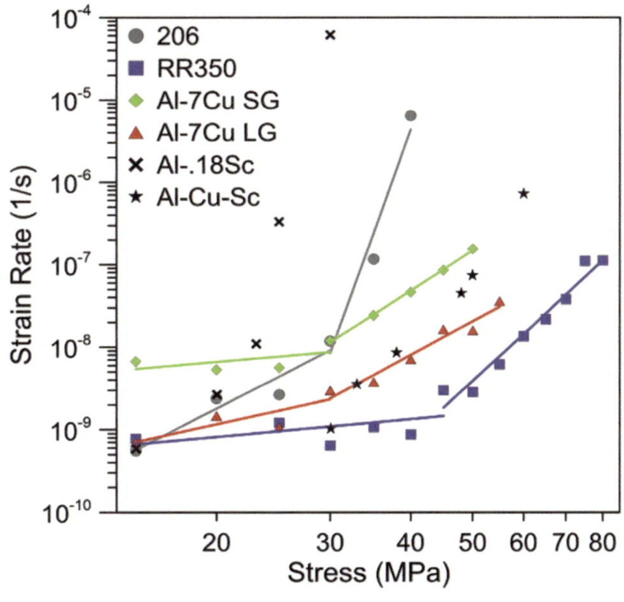

Figure 8.
Creep curves showing the steady state creep strain rate as a function of applied stress for various Al-Cu-Mn-Zr alloys (RR350, Al-7Cu SG and Al-7Cu LG where SG and LG refers to small and large grains, respectively) plus base Al-Cu and Al-Sc alloys [84].

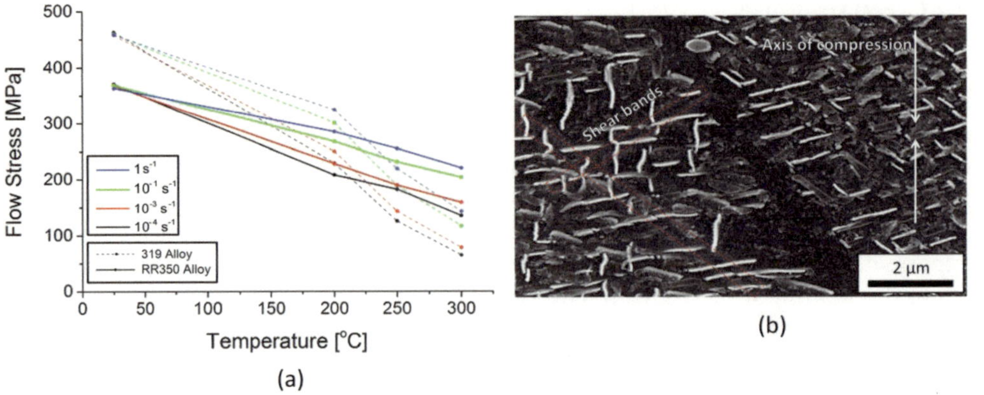

Figure 9.
(a) Comparison of flow stress variations with test temperatures for 319 and RR350 alloys at different strain rates, and (b) post-compression (at 300°C and 1 s^{-1} strain rate) SEM micrograph of RR350 alloy showing shear band formation and bending of θ′ precipitates within α-Al grains having <100> direction nearly parallel to the compression axis [87].

condition by conducting isothermal hot compression tests at different temperature-true strain rate combinations. At all strain rates, compressive flow stress of 319 alloy is greater than that for RR350 alloy up to 200°C (**Figure 9a**). However, within 250–300°C, RR350 alloy possesses higher flow stress which can be attributed to the stability of strengthening θ′ precipitates. In this temperature range, 319 alloy losses its flow stress by 40%. At lower strain rates (e.g. 10^{-4} s^{-1} & 10^{-3} s^{-1}), primary deformation mechanism for both alloys is strain hardening at room temperature, which changes to dynamic recovery and cross slip of dislocation at 250°C. Afterwards, dynamic recrystallization becomes predominant at 300°C while grain boundary sliding is the primary deformation mechanism at 350°C. In addition, RR350 alloy shows formation of shear bands as well as bending of θ′ precipitates within α-Al grains having <100> direction nearly parallel to the compression axis in the microstructure of specimens deformed at 300°C (**Figure 9b**).

10. Fatigue response

The excellent high temperature stability of Al-Cu-Mn-Zr alloys also make them a prime candidate for fracture critical engineering applications where fatigue properties are crucial consideration. Bahl et al. [88] studied the effect of Cu concentration on the high temperature (250°C) low cycle fatigue (LCF) properties of Al-Cu-Mn-Zr alloys that led to a correlation between LCF life and monotonic tensile fracture strain. At low strain amplitude (0.1%), Al-Cu-Mn-Zr alloys with either 6 wt% or 9 wt% of Cu do not undergo failure even after 10^5 number of cycles. However, at higher strain amplitudes (0.2% and 0.3%), the alloys fail within these many cycles of testing. This suggests that the fatigue life of Al-Cu-Mn-Zr alloys decreases with increasing strain amplitude.

For the peak-aged Al-Cu-Mn-Zr alloys with varying Cu content that underwent a further thermal exposure at 250°C for 100 hours, initiation of fatigue cracks almost always occur from the surface pores rather than from coarse grain boundary precipitates [80]. The fracture surfaces correspondingly do not contain much of the traces of intermetallic precipitates. Finite element modeling (FEM) also indicates that stress concentration at the pores are higher compared to that at the grain boundary precipitates. The low strain amplitude in the fatigue testing possibly led to the pore assisted crack initiation since otherwise the cracking from grain boundary precipitates would require higher stress concentration and their decohesion from the matrix which is only possible at larger strains [88].

Furthermore, since the variation in Cu content only affects the volume fraction of grain boundary intermetallic precipitates, it does not influence the cracking and in turn, low cycle fatigue behavior of Al-Cu-Mn-Zr alloys [80]. The thermal stability of θ′ precipitates also does not influence the fatigue property of these alloys since both crack initiation and propagation occur at a larger microstructural scale (from surface pores). It therefore appears that controlling the casting defects (predominantly shrinkage pores) is the most crucial factor to enhance the LCF life for Al-Cu-Mn-Zr alloys. Overall, these alloys exhibited moderate to excellent high temperature low cycle fatigue life making them suitable for components meant for elevated temperature applications.

11. Industrial application potentials of Al-Cu-Mn-Zr alloys

Industrial application of Al-Cu-Mn-Zr alloys require certain additional considerations on and above their excellent thermal stability and associated improvement in most of the high temperature mechanical properties as mentioned above. These include, but certainly not restricted to the ease of casting and defect formation, assessment of mechanical properties e.g. fatigue testing at larger component scale, possibility to adopt alternate component fabrication methodologies like additive manufacturing, wrought processing etc. [89]. Some of these aspects are mentioned below:

11.1 Hot tearing resistance

Hot tearing is a crucial casting defect that can affect the structural stability and properties of as-cast components [90, 91]. During solidification, molten metal usually remains in semi-solid state (mushy zone) for considerable duration. It also undergoes severe volume contraction and associated thermal stresses within the semi-solid metal regions. Under this condition, cracks form in the solidified component if there is an inadequate supply of molten mass to fill up the shrinkage volume. Controlling such defects in castings is difficult but extremely important for improving the fatigue life per se [91].

Sabau et al. [92] studied the hot tearing resistance of cast Al-Cu-Mn-Zr alloys with and without grain refiners in comparison to base Al-Cu and Al-Si alloys. The base Al-Cu alloy with >7 wt% Cu exhibits a grain refined microstructure in the casting. It also shows a decrease in the length of columnar to equiaxed transition zone that in turn improves the hot tearing resistance. In addition, simultaneous presence of Si and Fe (>0.2 wt%) increases the hot tearing resistance for this alloy. The low Cu containing alloys, on the other hand, possess coarse columnar grains within the cast microstructures which facilitates hot tearing for them. For the Al-Cu-Mn-Zr alloys, when Cu is added above 7 wt%, significant grain refinement occurs which further contributes to their excellent hot tearing resistance. In addition, when 0.1 wt% Ti is added as additional grain refiner, resultant Al-Cu-Mn-Zr alloy exhibits the finest microstructure and correspondingly, the best hot tearing resistance [89]. Ti added Al-Cu-Mn-Zr alloy was therefore speculated suitable for industrial applications [92].

In the measurement of hot tearing resistance, Sabau et al. [92] used an in-house multi-arm casting setup with varying arm length in a permanent mold. In this six-armed mold, the shortest arms were free from any visible cracks for all the alloys having varying amount of Cu and Ti, while the longest arm had severe cracking for these alloys. As per the visible inspection, a cracking index (C_i) was assigned to each of the arms where 0 corresponds to no crack condition and 2.5, 5, 7.5, and 10 was assigned to small, moderate, severe and completely fractured arms, respectively. Hot tearing index (HTI), M_s was then defined as the simple average of C_i over these six arms i.e.

$$M_s = \frac{1}{6}\sum_{i=1}^{6} C_i \tag{7}$$

The length of the arms, however, also plays a crucial role in the formation of these cracks. The longer arms are more susceptible to cracking compared to the shorter ones so that the weighted average was preferred for the calculation of HTI. The weighted average ($M_{J\text{-}K}$) was defined as:

$$M_{J-K} = \frac{\sum_{i=J}^{K} w_i C_i}{\sum_{i=J}^{K} w_i} \tag{8}$$

where, J and K corresponds to the shortest and longest arms and w_i is the weight factor which is inversely proportional to the length of the arm; J was considered as 2 and K as 5. As per the industrial standard, hot tearing index (HTI) equal or less than 3.3 is considered acceptable when C_i is in the range of 0–10. The HTI for various Al-Cu-Mn-Zr alloys with promising industrial application potential are listed in **Table 1**. Important to note that the fluidity of Al-Cu-Mn-Zr alloys with respect to temperature, especially under solidification conditions is also of immense importance for successful industrial application standpoint. However, no such data reporting the fluidity of Al-Cu-Mn-Zr alloys is available yet.

11.2 Additive manufacturing of Al-Cu-Mn-Zr alloy

In recent times, additive manufacturing (AM) is proven to be an extremely useful and alternate technique for shaping intricate parts in industrially relevant scales with excellent property combinations compared to cast counterparts [93]. For Al alloys however, additive manufacturing is a rather complicated and challenging process due to several factors like poor powder flowability, high thermal conductivity, laser reflectivity etc. [94]. In this regard, Shyam et al. [95] successfully fabricated AM parts from Al-Cu-Mn-Zr alloys by selective laser melting (SLM)

Alloy	Ti content (wt%)	M_{J-K}
Al-Cu-Mn-Zr with 8 wt% Cu	0.02	2.6
	0.1	2.01
	0.2	2.36
Al-Cu-Mn-Zr with 6.2 wt% Cu	0.1	2.6
	0.2	3.21
Al-Cu-Mn-Zr with 7.3 wt% Cu	0.1	2.49
	0.2	2.84

Table 1.
Hot tearing index (HTI) of different Al-Cu-Mn-Zr alloys with varying Ti content.

without any hot tearing using optimized processing parameters. The substrate temperature for SLM was kept 200°C which was sufficient for in-situ formation of strengthening θ′ precipitates.

Due to laser melting, AM Al-Cu-Mn-Zr alloys form typical "peacock tail" microstructure having overlapping melt pools. In addition, the AM microstructure consists of long columnar grains at the top and fine equiaxed grains at the bottom of the melt pools. Such refined AM microstructure yields comparatively higher strength up to 300°C compared to the cast Al-Cu-Mn-Zr alloys. The bimodal grain size distribution and refined grain boundary intermetallic precipitates further enhance the tensile elongation for AM Al-Cu-Mn-Zr alloys. However, creep properties for these AM alloys are somewhat compromised compared to the cast counterparts due to high proportion of grain boundaries in the refined AM microstructure. Overall, AM Al-Cu-Mn-Zr alloys are envisioned having potential in complex component manufacturing for high temperature applications owing to the simultaneous positive effects of refined microstructure and in situ formation of thermally stable strengthening θ′ precipitates [95].

11.3 Environmental impact from Al-Cu-Mn-Zr alloy

In the current global scenario, any new alloy development must help in reducing environmental impact e.g. carbon footprint and green house emissions [96]. The primary target area for Al-Cu-Mn-Zr alloys is automotive industry, which also formed the early motivation of their inception and further development [97]. The aim was to develop Al alloys for engine components that experience high working temperature (~300°C) e.g. cylinder heads in combustion engines. The use of Al-Cu-Mn-Zr alloys for making such components can effectively raise the working temperature and increase the fuel efficiency of next generation passenger vehicles, thereby proving environment friendly in terms of fuel consumption.

12. Conclusion

The present book chapter elucidates a comprehensive review about the development as well as the science and technology behind the new-age Al-Cu-Mn-Zr (ACMZ) alloys. The major observations are summarized below.

- The new age Al-Cu-Mn-Zr (ACMZ) alloy developed in recent times by micro-alloying conventional Al-Cu alloys with Mn and Zr shows unprecedented microstructural stability up to ~350°C.

- The strengthening θ′ precipitates in Al-Cu-Mn-Zr alloys are stabilized primarily due to preferential solute segregation (Mn & Zr) at the θ′ precipitate/α-Al matrix interfaces which helps in reducing their interfacial energy. Mn atoms segregate at the mobile semi-coherent interfaces while Zr atoms primarily sits at the corner of coherent/semi-coherent interfaces.

- The difference in the diffusivity of micro-alloying elements decides the sequence of their segregation at θ′ precipitate/α-Al matrix interfaces. This in turn plays a synergistic role in stabilizing θ′ precipitate at different temperature ranges. Faster diffusing Mn initially segregates at both coherent and semi-coherent interfaces and provides thermal stability at lower temperatures. Zr being relatively slower diffusing element segregates at a later stage but effectively restricts θ′ precipitates from transforming to stable θ precipitates up to a much higher temperature.

- Various other mechanisms e.g. solute drag, diffusion barrier formation, ledge poisoning by co-precipitation of thermally stable intermetallics etc. also contributes to the coarsening resistance of θ′ precipitates.

- Trace elements like Ti and Si further influence θ′ precipitate stability in Al-Cu-Mn-Zr alloys. Si above a critical content (>0.1 wt%) potentially substitutes Mn and Zr at the θ′ precipitate/α-Al matrix interfaces, thereby decreasing the stability of θ′ precipitates at elevated temperatures. On the other hand, $L1_2$ structured $Al_3(Zr_xTi_{1-x})$ precipitates form at θ′ precipitate interfaces in presence of Ti which further adds to the stability of these alloys at higher temperatures.

- DFT simulations confirm the reduction of interfacial energy from segregation of micro-alloying elements and provide the segregation profiles at various (coherent and semi-coherent) interfaces. Phase field simulations further suggest that the cumulative effect of interfacial energy reduction and solute drag led to the θ′ precipitate stabilization at 300°C. At 400°C, reduction in the mobility of Cu atoms is required in addition to reduced interfacial energy for efficient resistance towards θ′ precipitate coarsening.

- As a result of the excellent thermal stability of strengthening θ′ precipitates, Al-Cu-Mn-Zr alloys exhibit superior high temperature tensile properties (hardness, yield and ultimate tensile strengths etc.) compared to base Al-Cu or Al-Si-Cu alloys. For example, Al-Cu-Mn-Zr alloys possess almost twice the yield strength compared to base Al-Cu alloy after prolonged thermal exposure at 300°C.

- At high temperature (300–350°C), compressive flow stress of Al-Cu-Mn-Zr alloy (RR350) is higher than conventional Al-Si-Cu (319) alloy. The high temperature deformation mechanisms for the former alloy vary from strain hardening at room temperature to cross slip and dynamic recovery as the test temperature increases at 250°C. Further higher test temperature promotes dynamic recrystallization of the matrix at 300°C and finally lead to grain boundary sliding to 350°C.

- The Al-Cu-Mn-Zr alloy shows excellent creep resistance compared to the base Al-Cu alloys. For the former alloy, diffusional Coble creep is the controlling mechanism at low stress levels which switches to dislocation creep at higher stress levels.

- In high temperature low cycle fatigue testing, Al-Cu-Mn-Zr alloy does not fail up to 10^5 cycles at low stress amplitudes whereas the fatigue life decreases with increasing stress amplitude. The casting defects (pores) are found to be more influential factor by means of stress concentration and crack initiation than either the stability of θ′ precipitates or the presence of grain boundary intermetallic precipitates.

- The Al-Cu-Mn-Zr alloy exhibits superior hot tearing resistance compared to the conventional Al-Cu alloys. Additive manufacturing of Al-Cu-Mn-Zr alloy also shows promising property combinations owing to a refined microstructure and in-situ formation of strengthening θ′ precipitates. In addition to the positive environmental impacts, both of these factors are crucial for their suitability in industrial production purpose.

Overall, Al-Cu-Mn-Zr alloys offer huge potential for industrial applications where lightweight materials are sought after for excellent high temperature mechanical properties. These new-age alloys can in fact prove to be a game changer for the existing passenger vehicle engines and may bring a paradigm shift in the automobile sectors. Owing to low density and excellent high temperature stability, their future use is certainly leading to a safer and greener environment.

Author details

Samarendra Roy[1] and Shibayan Roy[2]*

1 School of Nano Science and Technology, Indian Institute of Technology, Kharagpur, India

2 Materials Science Centre, Indian Institute of Technology, Kharagpur, India

*Address all correspondence to: shibayan@matsc.iitkgp.ac.in

IntechOpen

© 2022 The Author(s). Licensee IntechOpen. This chapter is distributed under the terms of the Creative Commons Attribution License (http://creativecommons.org/licenses/by/3.0), which permits unrestricted use, distribution, and reproduction in any medium, provided the original work is properly cited.

References

[1] Mazzolani F. Aluminium alloy structures. London, UK: CRC Press; 1994. Available from: https://www.taylorfrancis.com/books/mono/10.1201/9781482267198/aluminium-alloy-structures-federico-mazzolani

[2] Heinz A et al. Recent development in aluminium alloys for aerospace applications. Materials Science and Engineering A. 2000;**280**(1):102-107

[3] Campbell FC. Elements of Metallurgy and Engineering Alloys. Ohio, USA: ASM International; 2008

[4] Hirsch J. Recent development in aluminium for automotive applications. Transactions of Nonferrous Metals Society of China. 2014;**24**(7):1995-2002

[5] Mrazova M. Advanced composite materials of the future in aerospace industry. Incas bulletin. 2013;**5**(3):139

[6] Nappi C. The global aluminium industry 40 years from 1972. World Aluminium. 2013:1-27

[7] Murray JL. The aluminum-copper system. International Metals Reviews. 1985;**30**:211-233

[8] Martin JW. Precipitation Hardening: Theory and Applications. Oxford, UK: Butterworth-Heinemann; 2012

[9] Polmear IJ. Aluminium Alloys–A Century of Age Hardening. Materials forum. 2004;**28**:1-14

[10] Wang S, Fan C. Crystal structures of Al2Cu revisited: Understanding existing phases and exploring other potential phases. Metals. 2019;**9**(10):1037

[11] Shen Z et al. Atomic-scale mechanism of the $\theta'' \rightarrow \theta'$ phase transformation in Al-Cu alloys. Journal of Materials Science and Technology. 2017;**33**(10):1159-1164

[12] Porter DA, Easterling KE. Phase Transformations in Metals and Alloys (Revised Reprint). Boca Raton, Florida, USA: CRC press; 2009

[13] Chisholm MF et al. Atomic structures of interfacial solute gateways to θ' precipitates in Al-Cu alloys. Acta Materialia. 2021;**212**:116891

[14] Martin JW. Precipitation Hardening. 2nd ed. Oxford, UK: Butterworth-Heinemann; 1998

[15] Roy S et al. Comparative evaluation of cast Aluminum alloys for automotive cylinder heads: Part II—Mechanical and thermal properties. Metallurgical and Materials Transactions A. 2017;**48**(5):2543-2562

[16] Orowan E. Theory of yield without particle shear, symposium on internal stresses in metals and alloys. In: Martin JW, editor. Precipitation Hardening. Vol. 1968. Oxford: Pergamon Press; 1948

[17] Nie JF, Muddle BC, Polmear IJ. The effect of precipitate shape and orientation on dispersion strengthening in high strength aluminium alloys. Materials Science Forum. 1996;**217-222**:1257-1262

[18] Hornbogen E. Hundred years of precipitation hardening. Journal of Light Metals. 2001;**1**:127-132

[19] Gleiter H, Hornbogen E. Precipitation hardening by coherent particles. Materials Science and Engineering. 1967;**2**:285-302

[20] Guan R et al. A high-strength, ductile Al-0.35 Sc-0.2 Zr alloy with good electrical conductivity strengthened by coherent nanosized-precipitates. Journal of Materials Science and Technology. 2017;**33**(3):215-223

[21] Tiryakioğlu M, Campbell J, Staley J. The influence of structural integrity on the tensile deformation of cast Al–7wt.% Si–0.6 wt.% Mg alloys. Scripta Materialia. 2003;**49**(9):873-878

[22] Roy S, Allard LF, Rodriguez A, Watkins TR, Shyam A. Comparative evaluation of cast aluminum alloys for automotive cylinder heads: part I–microstructure evolution. Metallurgical and Materials Transactions A. May 2017; **48**(5):2529-2542

[23] Vaithyanathan V, Wolverton C, Chen L. Multiscale modeling of θ′ precipitation in Al–Cu binary alloys. Acta Materialia. 2004;**52**(10):2973-2987

[24] Dahmen U, Westmacott K. Ledge structure and the mechanism of θ′ precipitate growth in Al Cu. Physica Status Solidi. 1983;**80**(1): 249-262

[25] Li H et al. Precipitation evolution and coarsening resistance at 400 C of Al microalloyed with Zr and Er. Scripta Materialia. 2012;**67**(1):73-76

[26] Booth-Morrison C, Dunand DC, Seidman DN. Coarsening resistance at 400 C of precipitation-strengthened Al–Zr–Sc–Er alloys. Acta Materialia. 2011;**59**(18):7029-7042

[27] Wen S et al. Precipitation evolution in Al–Er–Zr alloys during aging at elevated temperature. Journal of Alloys and Compounds. 2013;**574**: 92-97

[28] Knipling KE, Dunand DC. Creep resistance of cast and aged Al–0.1 Zr and Al–0.1 Zr–0.1 Ti (at.%) alloys at 300–400 C. Scripta Materialia. 2008;**59**(4): 387-390

[29] Peng G et al. A study of nanoscale Al3 (Zr, Yb) dispersoids structure and thermal stability in Al–Zr–Yb alloy. Materials Science and Engineering A. 2012;**535**:311-315

[30] Wang M et al. Low-cycle fatigue properties and life prediction of Al-Si piston alloy at elevated temperature. Materials Science and Engineering A. 2017;**704**:480-492

[31] Fan K et al. Elevated temperature low cycle fatigue of a gravity casting Al–Si–Cu alloy used for engine cylinder heads. Materials Science and Engineering A. 2015;**632**:127-136

[32] Gao T et al. Evolution, microhardness of ZrAlSi intermetallic and its impact on the elevated-temperature properties in Al–Si alloys. Materials Science and Engineering A. 2012;**552**:523-529

[33] Gao T et al. Morphological evolution of ZrAlSi phase and its impact on the elevated-temperature properties of Al–Si piston alloy. Journal of Alloys and Compounds. 2013;**567**:82-88

[34] Colombo M, Gariboldi E, Morri A. Influences of different Zr additions on the microstructure, room and high temperature mechanical properties of an Al-7Si-0.4 Mg alloy modified with 0.25% Er. Materials Science and Engineering A. 2018;**713**:151-160

[35] Kasprzak W, Amirkhiz BS, Niewczas M. Structure and properties of cast Al–Si based alloy with Zr–V–Ti additions and its evaluation of high temperature performance. Journal of Alloys and Compounds. 2014;**595**: 67-79

[36] Hernandez-Sandoval J et al. The ambient and high temperature deformation behavior of Al–Si–Cu–Mg alloy with minor Ti, Zr, Ni additions. Materials & Design. 2014;**58**:89-101

[37] Zuo L et al. Effect of ε-Al3Ni phase on mechanical properties of Al–Si–Cu–Mg–Ni alloys at elevated temperature. Materials Science and Engineering A. 2020;**772**:138794

[38] Jeon J, Shin J, Bae D. Si phase modification on the elevated temperature mechanical properties of Al-Si hypereutectic alloys. Materials Science and Engineering A. 2019;**748**: 367-370

[39] Chen K et al. The role of various Zr additions in static softening behavior of Al-Zn-Mg-Cu alloys during interval holding of double-stage hot deformation. Journal of Alloys and Compounds. 2019;**792**:1112-1121

[40] Yao D et al. Effects of La addition on the elevated temperature properties of the casting Al–Cu alloy. Materials Science and Engineering A. 2011;**528**(3): 1463-1466

[41] Balducci E et al. Thermal stability of the lightweight 2099 Al-Cu-Li alloy: Tensile tests and microstructural investigations after overaging. Materials & Design. 2017;**119**:54-64

[42] Rao KS et al. Microstructure and high temperature strength of age hardenable AA2219 aluminium alloy modified by Sc, Mg and Zr additions. Materials Science and Technology. 2009;**25**(1):92-101

[43] Gao Y et al. Stabilizing nanoprecipitates in Al-Cu alloys for creep resistance at 300°C. Materials Research Letters. 2019;**7**(1):18-25

[44] Gao Y et al. Co-stabilization of θ′-Al2Cu and Al3Sc precipitates in Sc-microalloyed Al–Cu alloy with enhanced creep resistance. Materials Today Nano. 2019;**6**:100035

[45] Makineni SK et al. Enhancing elevated temperature strength of copper containing aluminium alloys by forming L1 2 Al 3 Zr precipitates and nucleating θ″ precipitates on them. Scientific Reports. 2017;**7**(1):1-9

[46] Rakhmonov J et al. Enhanced mechanical properties of high-temperature-resistant Al–Cu cast alloy by microalloying with Mg. Journal of Alloys and Compounds. 2020;**827**:154305

[47] Song M, Xiao D, Zhang F. Effect of Ce on the thermal stability of the Ω phase in an Al-Cu-Mg-Ag alloy. Rare Metals. 2009;**28**(2):156-159

[48] Shyam A et al. Elevated temperature microstructural stability in cast AlCuMnZr alloys through solute segregation. Materials Science and Engineering A. 2019;**765**:138279

[49] Polmear IJ. Role of Trace Elements in Aged Aluminium-Alloys. In: Materials Science Forum. Trans Tech Publications Ltd. 1987;**13**:195-214

[50] Silcock J, Flower H. Comments on a comparison of early and recent work on the effect of trace additions of Cd, In, or Sn on nucleation and growth of θ′ in Al–Cu alloys. Scripta Materialia. 2002; **46**(5):389-394

[51] Ringer S, Hono K, Sakurai T. The effect of trace additions of sn on precipitation in Al-Cu alloys: An atom probe field ion microscopy study. Metallurgical and Materials Transactions A. 1995;**26**(9):2207-2217

[52] Bahl S et al. Aging behavior and strengthening mechanisms of coarsening resistant metastable θ' precipitates in an Al–Cu alloy. Materials & Design. 2021;**198**:109378

[53] Shin D et al. Solute segregation at the Al/θ′-Al2Cu interface in Al-Cu alloys. Acta Materialia. 2017;**141**: 327-340

[54] Samolyuk GD et al. Equilibrium solute segregation to matrix-θ′ precipitate interfaces in Al-Cu alloys from first principles. Physical Review Materials. 2020;**4**(7):073801

[55] Ling J et al. Machine learning for alloy composition and process

optimization. In: Proc. of Turbo Expo: Power for Land, Sea, and Air. New York, USA: American Society of Mechanical Engineers; 2018

[56] Li J et al. High-throughput simulation combined machine learning search for optimum elemental composition in medium entropy alloy. Journal of Materials Science and Technology. 2021;**68**:70-75

[57] Horstemeyer MF. Integrated Computational Materials Engineering (ICME) for Metals: Using Multiscale Modeling to Invigorate Engineering Design with Science. Hoboken, New Jersey, USA: John Wiley & Sons; 2012

[58] Sabau AS et al. Process simulation role in the development of new alloys based on an integrated computational materials engineering approach. In: Proc. of ASME International Mechanical Engineering Congress and Exposition. USA: American Society of Mechanical Engineers; 2014

[59] Shyam A et al. High Performance Cast Aluminum Alloys for Next Generation Passenger Vehicle Engines. Oak Ridge, TN (United States): Oak Ridge National Lab.(ORNL); 2018

[60] Wang WY et al. Integrated computational materials engineering for advanced materials: A brief review. Computational Materials Science. 2019; **158**:42-48

[61] Shin D et al. Petascale supercomputing to accelerate the design of high-temperature alloys. Science and Technology of Advanced Materials. 2017;**18**(1):828-838

[62] Poplawsky JD et al. The synergistic role of Mn and Zr/Ti in producing θ′/L12 co-precipitates in Al-Cu alloys. Acta Materialia. 2020;**194**:577-586

[63] Shower P et al. The role of Si in determining the stability of the θ′ precipitate in Al-Cu-Mn-Zr alloys. Journal of Alloys and Compounds. 2021; **862**:158152

[64] Wang J et al. Structural deformation and transformation of θ′-Al2Cu precipitate in Al matrix via interfacial diffusion. Computational Materials Science. 2019;**156**:111-120

[65] Rummel G et al. Diffusion of implanted 3d-transition elements in aluminium part I: Temperature dependence/diffusion implantierter 3d-Übergangselemente in aluminium Teil I: Temperaturabhängigkeit. International Journal of Materials Research. 1995; **86**(2):122-130

[66] Kidson G, Miller G. A study of the interdiffusion of aluminum and zirconium. Journal of Nuclear Materials. 1964;**12**(1):61-69

[67] Lifshitz IM, Slyozov VV. The kinetics of precipitation from supersaturated solid solutions. Journal of Physics and Chemistry of Solids. 1961;**19**(1–2):35-50

[68] Boyd J, Nicholson R. The coarsening behaviour of θ ″and θ′ precipitates in two Al-Cu alloys. Acta Metallurgica. 1971;**19**(12):1379-1391

[69] Mitlin D, Morris J, Radmilovic V. Catalyzed precipitation in Al-Cu-Si. Metallurgical and Materials Transactions A. 2000;**31**(11):2697-2711

[70] Bonfield W, Datta P. Precipitation hardening in an Al-Cu-Si-Mg alloy at 130 to 220° C. Journal of Materials Science. 1976;**11**(9):1661-1666

[71] Yoshida S, Kiritani M, Shimomura Y. Dislocation loops with stacking fault in quenched aluminum. Journal of the Physical Society of Japan. 1963;**18**(2):175-183

[72] Thomas G, Whelan M. Helical dislocations in quenched aluminium-4%

copper alloys. Philosophical Magazine. 1959;**4**(40):511-527

[73] Hutchinson CR, Ringer SP. Precipitation processes in Al-Cu-Mg alloys microalloyed with Si. Metallurgical and Materials Transactions A. 2000; **31**(11):2721-2733

[74] Bonfield W, Datta P. Zone formation during room temperature ageing of Al-4% Cu-0.8% Si-0.8% Mg. Journal of Materials Science. 1977;**12**(5):1050-1052

[75] Srinivasan S, Desch PB, Schwarz RB. Metastable phases in the Al sub 3 X (X = Ti, Zr, and Hf) intermetallic system. Scripta Metallurgica; (United States), 1991;**25**(11). p. Medium: X; Size: Pages: 2513-2516 2009-12-16

[76] Shower P et al. Mechanisms for stabilizing θ′(Al2Cu) precipitates at elevated temperatures investigated with phase field modeling. Materialia. 2019;**6**: 100335

[77] Kim K et al. First-principles/phase-field modeling of θ′ precipitation in Al-Cu alloys. Acta Materialia. 2017;**140**: 344-354

[78] Fine M et al. Basic principles for selecting phases for high temperature metal matrix composites: Interfacial considerations. Scripta Metallurgica. 1988;**22**(6):907-910

[79] Shower P et al. Temperature-dependent stability of θ′-Al2Cu precipitates investigated with phase field simulations and experiments. Materialia. 2019;**5**:100185

[80] Bahl S et al. Effect of copper content on the tensile elongation of Al–Cu–Mn–Zr alloys: Experiments and finite element simulations. Materials Science and Engineering A. 2020;**772**:138801

[81] da Costa Teixeira J et al. The effect of shear-resistant, plate-shaped precipitates on the work hardening of Al alloys: Towards a prediction of the strength–elongation correlation. Acta Materialia. 2009;**57**(20):6075-6089

[82] Simar A et al. Sequential modeling of local precipitation, strength and strain hardening in friction stir welds of an aluminum alloy 6005A-T6. Acta Materialia. 2007;**55**(18):6133-6143

[83] Ardell AJ. Precipitation hardening. Metallurgical Transactions A. 1985; **16**(12):2131-2165

[84] Milligan BK et al. Impact of microstructural stability on the creep behavior of cast Al–Cu alloys. Materials Science and Engineering A. 2020;**772**: 138697

[85] Sokolowski JH et al. Improvement of 319 aluminum alloy casting durability by high temperature solution treatment. Journal of Materials Processing Technology. 2001;**109**(1–2):174-180

[86] Singer R, Blum W, Ilschner B. Deformation-induced microstructural instability in a θ′-hardened aluminum alloy at high temperature. Materials Science and Engineering. 1979;**40**(2): 235-243

[87] Shower P et al. The effects of microstructural stability on the compressive response of two cast aluminum alloys up to 300° C. Materials Science and Engineering A. 2017;**700**: 519-529

[88] Bahl S et al. Influence of copper content on the high temperature tensile and low cycle fatigue behavior of cast Al-Cu-Mn-Zr alloys. International Journal of Fatigue. 2020;**140**:105836

[89] Sabau AS et al. Grain refinement effect on the hot-tearing resistance of higher-temperature Al–Cu–Mn–Zr alloys. Metals. 2020;**10**(4):430

[90] Sigworth G. Hot tearing of metals. Transactions of the American

Foundrymen's Society. 1996;**106**: 1053-1062

[91] Li Y et al. Recent advances in hot tearing during casting of aluminium alloys. Progress in Materials Science. 2021;**117**:100741

[92] Sabau AS et al. Hot-tearing assessment of multicomponent nongrain-refined Al-Cu alloys for permanent Mold castings based on load measurements in a constrained Mold. Metallurgical and Materials Transactions B. 2018;**49**(3):1267-1287

[93] Frazier WE. Metal additive manufacturing: A review. Journal of Materials Engineering and Performance. 2014;**23**(6):1917-1928

[94] Aboulkhair NT et al. 3D printing of aluminium alloys: Additive manufacturing of aluminium alloys using selective laser melting. Progress in Materials Science. 2019;**106**:100578

[95] Shyam A et al. An additively manufactured AlCuMnZr alloy microstructure and tensile mechanical properties. Materialia. 2020;**12**:100758

[96] Kumar D, Phanden RK, Thakur L. A review on environment friendly and lightweight magnesium-based metal matrix composites and alloys. Materials Today: Proceedings. 2021;**38**:359-364

[97] Roy S, Allard LF, Rodriguez A, Porter WD, Shyam A. Comparative evaluation of cast aluminum alloys for automotive cylinder heads: Part II– mechanical and thermal properties. Metallurgical and Materials Transactions A. 2017;**48**(5):2543-2562

Chapter 3

Characteristics of Al-Mg Test Pieces with Fe Impurities Fabricated by Die Casting, Roll Casting, and Hot Forging

Toshio Haga

Abstract

The suitability of Al-Mg alloys for recycling was investigated using energy-saving processes. The Al-Mg alloy is a non-heat-treatable alloy and has the advantage of energy saving in comparison with heat-treatable alloys. Al-Mg alloys with Mg contents ranging from 4.5–10% were tested. Die casting, cast-forging, and roll casting were selected as energy-saving processes, as they have the advantage of process saving. A single-roll caster equipped with a scraper was used as the roll-caster. Fe was added to the Al-Mg alloys at contents of 0.2%, 0.4%, 0.6%, and 0.8% to model recycled alloys used in automobile manufacture. In the selected processes, the tensile stress and 0.2% proof stress of the Al-Mg alloys were little influenced by the added Fe content, whereas the elongation tended to decrease as the Fe content increased. The process influenced the degree to which the Fe content affected the elongation, and it was found that a suitable Mg content for recycling depends on the target process.

Keywords: Al-Mg alloy, non-heat treatment aluminum alloy, Al-Mg alloy with Fe, die cast, cast-forging, roll-casting, recycle aluminum alloy

1. Introduction

Al-Mg alloys are non-heat-treatable aluminum alloys, which means they save energy in comparison with heat-treatable aluminum alloys. The Al-Si-Mg alloy, which is a heat-treatable aluminum alloy, is commonly used in automobile manufacture. For example, 6061 is used for forging, 6022 is used for sheet forming, and A357 and Silafont-37™ are used for casting and die casting. When aluminum alloys are used for automobile manufacture, Fe impurities are incorporated into the alloy, which causes AlSiFe intermetallic compounds to solidify when Al-Si-Mg alloys are recycled, reducing the Si content in the Al-Si-Mg alloy as AlSiFe intermetallic compound was crystallized. In Al-Mg-Si alloys, Mg_2Si precipitates during aging, causing the strength of the Al-Si-Mg alloy to increase. When the AlSiFe intermetallic compounds solidify, this can cause a shortage of Si for the Mg_2Si, and the strength may not increase sufficiently [1, 2]. As a result, Fe impurities have a reduced effect on Al-Mg alloys in comparison with Al-Si-Mg alloys. This is the second advantage of Al-Mg alloys over Al-Si-Mg heat-treatable alloys. Much less work has been done to

IntechOpen

investigate the effect of Fe impurities on the mechanical properties of Al-Mg alloys than for of Al-Si-Mg alloys [1–5].

To be suitable for the recycling of aluminum alloys, the selected process must satisfy the following two requirements: saving energy and improving the deterioration of the mechanical properties of the recycled alloy. In this paper, die casting, cast-forging, and roll casting were selected as the processes using the recycled Al-Mg alloys. Die casting can be used to produce aluminum alloy parts in one process with rapid solidification. Cast-forging has the advantage of energy-saving by process saving and the deformation effect, as the casting structure becomes the deformation structure [5, 6]. Roll casting has the advantage of energy-saving by process saving and rapid solidification. The intermetallic composition including Fe impurities becomes fine as a result of the rapid solidification.

In the recycling of aluminum alloys used for automobiles, the Fe content of the aluminum alloy is estimated to increase by 0.2% after shredding [7]. In this study, Fe contents of 0.2%, 0.4%, 0.6% and 0.8% were added to Al-Mg alloys to model recycled Al-Mg alloys. The addition of 0.8% Fe is considered to represent an alloy that has been recycled four times.

The Mg contents of the Al-Mg alloys used in this study were 4.5%, 6%, 8%, and 10%. The Mg content of 4.5% is near that of the 514.0 and 5182 aluminum alloys, and the Mg content of 8% is near that of 518.0. The four Fe contents were added to these four Al-Mg alloys, and test pieces were fabricated with the three selected processes. The mechanical properties were investigated via a tensile test. A deep drawing test was conducted on the plates made from the strips cast by the rolling caster. The suitability of the different Al-Mg alloys for recycling was then evaluated based on the obtained results.

2. Die casting

2.1 Die casting machine and die

A 500 kN cold chamber die casting machine (Hishinuma Machinery HC 50F) with an injection power of 100 kN and a sleeve diameter of 45 mm was used in this study. The plunger speed was 1.6 m/s. The test piece used for the tensile test and the spiral die used for the fluidity test are shown in **Figure 1**.

2.2 Fluidity test

The effects of the Mg and Fe contents on the fluidity were investigated. The results of the fluidity test are shown in **Figure 2**. The fluidity of the Al-Mg alloy

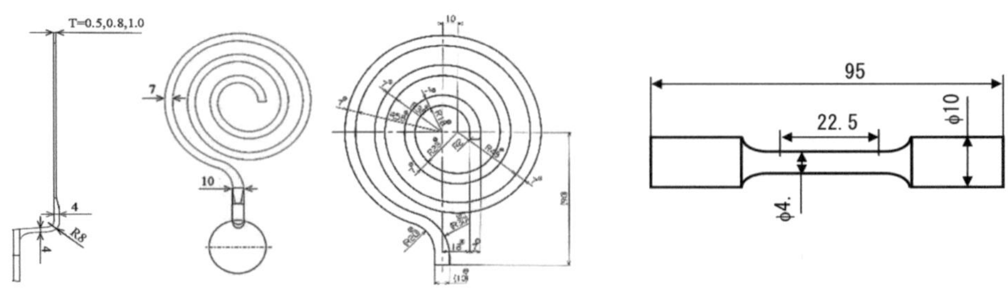

Figure 1.
Test piece for the spiral die for the fluidity test and the tensile test.

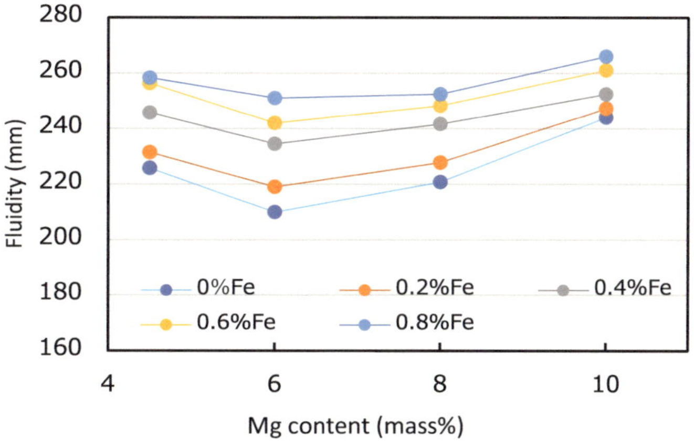

Figure 2.
Fluidity of Al-Mg alloys plotted against Mg content at different added Fe contents.

decreased with increasing Mg content until 6% Mg and then increased as the Mg content was increased beyond 6%. The fluidity was greatest at an Mg content of 10% and progressively decreased at contents of 4.5%, 8%, and 6%.

The fluidity increased with increasing Fe content, as shown in **Figure 2**. The flow stress at the semisolid condition decreased with increasing Fe content because the primary crystal became smaller and exhibited the mushy condition as the Fe content was increased. It is known that Fe is added to aluminum alloys during die casting to prevent the sticking of the solidification layer to the die. The heat transfer between the solidification layer and the die decreases with increasing Fe content because the contact condition between the solidification layer and the die worsens. As a result, the solidification time decreases and the fluidity increases. The increase of the Fe content during recycling; thus, does not make the fluidity worse but better.

2.3 Effect of Mg content on tensile test results

Figure 3 shows the results of the tensile test of the die-cast test pieces plotted against the Mg content. Both the tensile stress and the 0.2% proof stress gradually increased with increasing Mg content. The elongation was maximized at 6% Mg and remarkably decreased with further increases in the Mg content to 8% and 10% Mg. The elongation of the Al-6%Mg was 17.4%, which is excellent. The elongation of the Al-10%Mg was 4.7%, which is better than that of A383, a popular alloy for die casting in Japan. These Al-Mg alloys have 0.2% proof stresses and elongations that are better than those of A383. These results demonstrate that the Mg content should be selected based on the target user. If ductility is important, Al-6%Mg is better, whereas if strength is important, Al-8%Mg or Al-10%Mg is suitable.

2.4 Effect of added Fe content on tensile test results

The tensile stress, 0.2% proof stress, and elongation of different Al-Mg alloys are plotted against added Fe content in **Figures 4–6**, respectively. When Fe was added, the tensile stress of each Al-Mg alloy was the same as or better than that of the corresponding Al-Mg alloy without added Fe. In die casting using recycled

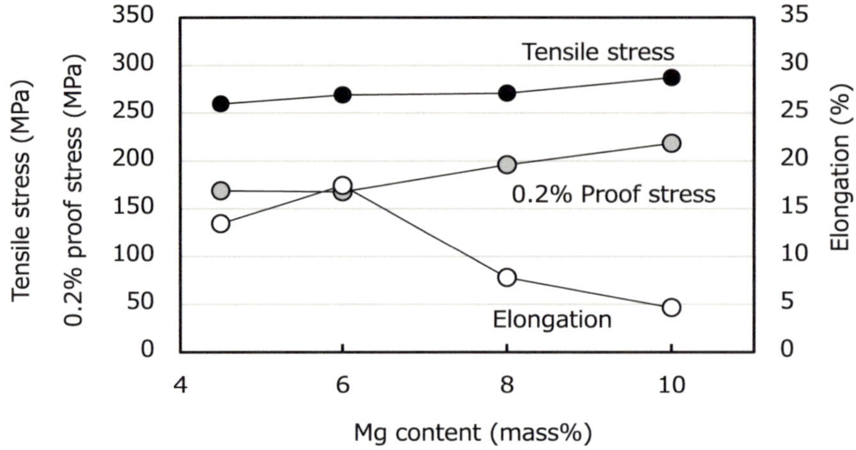

Figure 3.
Effect of the Mg content of the Al-Mg alloy on the result of the tensile test of the die-cast test piece.

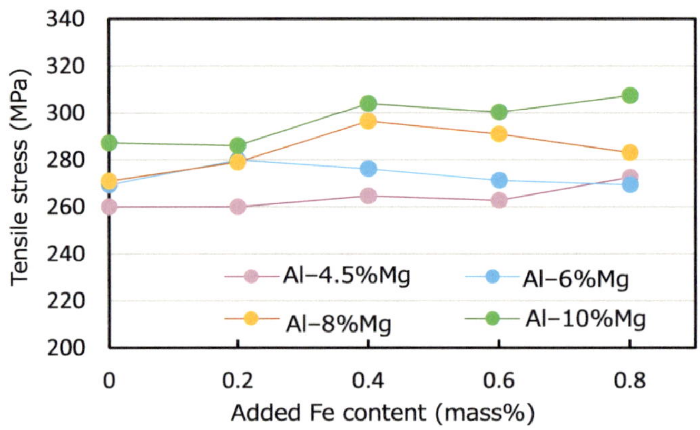

Figure 4.
Tensile stress of different die-cast Al-Mg alloys plotted against added Fe content.

Al-Mg alloys, it was clear that the tensile stress was not degraded by increasing Fe content.

The 0.2% proof stress of the Al-Mg alloys with different added Fe contents is shown in **Figure 5**. The results indicate that the 0.2% proof stress was not significantly affected by the addition of Fe. The 0.2% proof stress of the Al-8%Mg and Al-10%Mg increased with the addition of Fe.

The elongation of the Al-Mg alloys with different added Fe contents is shown in **Figure 6**. The amount of decrease in the elongation with increasing Fe content was dependent on the Mg content of the alloy. The elongation of the Al-4.5%Mg decreased substantially with the addition of 0.2% Fe but changed little with further increases in Fe content up to 0.8%. The elongation of the Al-4.5%Mg with 0.8%Fe was 10.2%. It was clear that the elongation of the Al-4.5%Mg was not greatly influenced by the Fe content for Fe contents above 0.2%. The reduction in the elongation from an Fe content of 0 to 0.2% was smaller at greater Mg contents. The elongation of the Al-6%Mg decreased almost linearly from 17.4–8% as the Fe content increased from 0.2% to 0.8%. At Fe contents of 0.2% and 0.4%, the elongation of the Al-6%Mg was greater than that of the Al-4.5%Mg. The elongations of the Al-8%Mg and the Al-10%Mg gradually decreased with increasing Fe content. The elongations of the Al-4.5%Mg, Al-6%Mg, and Al-8%Mg, each with 0.8%Fe, and the Al-10%Mg with 0.4%Fe were greater than

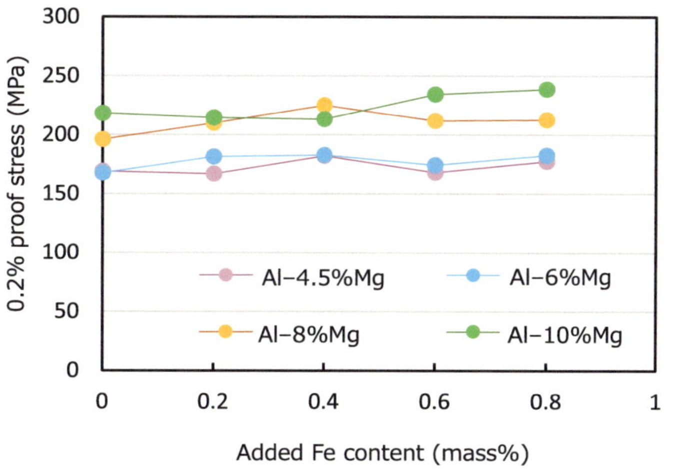

Figure 5.
The 0.2% proof stress of die-cast Al-Mg alloys plotted against added Fe content.

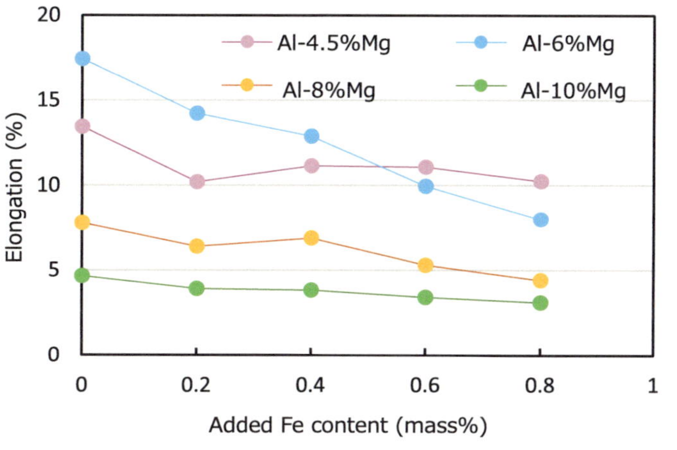

Figure 6.
Elongation of die cast Al-Mg alloys plotted against Fe content.

the elongation of the A383, which is 3.5%. This means that die-cast Al-Mg alloys may be suitable for recycling when die-cast Al-Mg alloys are used for automobile parts.

3. Cast-Forging

3.1 Model process of cast-forging

In cast-forging, a preform is cast near the net shape, which is suitable for forging. Processing and energy can be saved by cast-forging. The hot forging of a gravity-cast ingot was conducted as a model of cast-forging [8–10]. The process of gravity casting and hot forging is shown in **Figure 7**. This process is similar to cast-forging.

The cooling rate of the gravity-cast Al-4.5%Mg ingot was 30.6°C/s. A specimen was made for the tensile test and tested to investigate the mechanical properties of the ingot. A square bar was cut out from the as-cast ingot and heated at 500°C for 1 h. The forging was conducted at 50% reduction. The mechanical properties of the

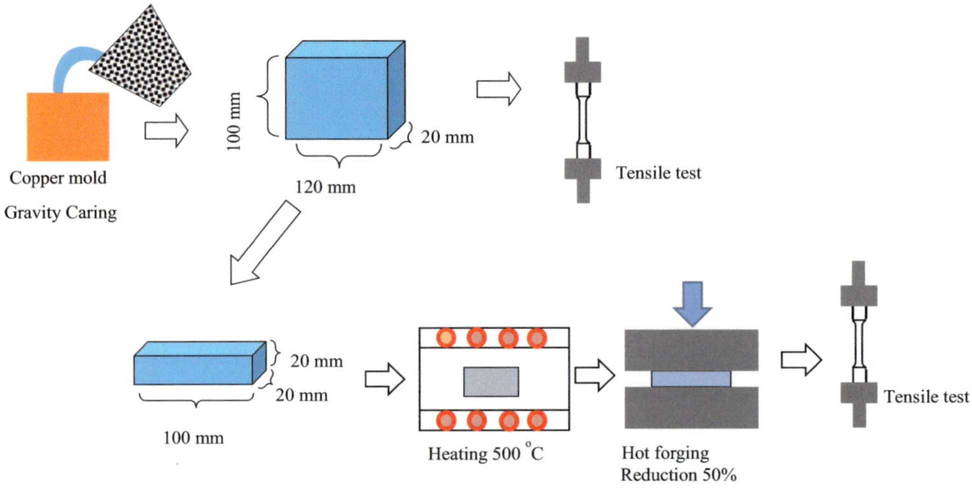

Figure 7.
Schematic of process from gravity casting to hot forging.

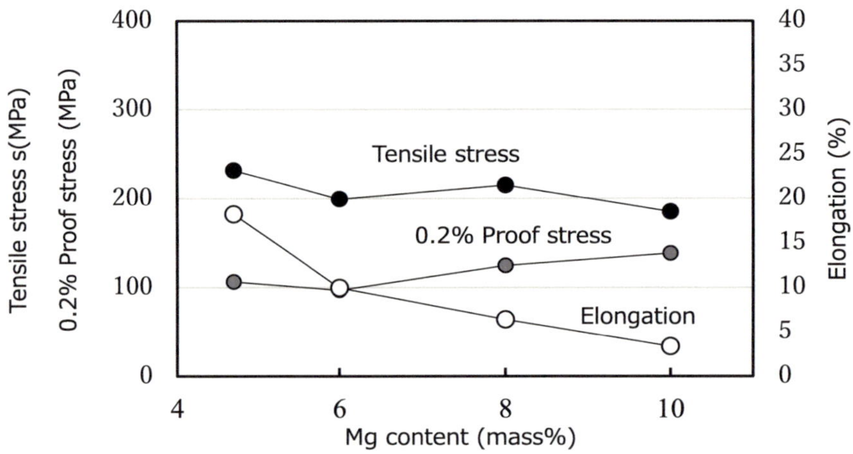

Figure 8.
Tensile test results for gravity-cast Al-Mg alloys.

hot-forged rectangular bar were investigated by tensile testing. The dimensions of the test piece are shown in **Figure 1a**.

3.2 Tensile test of gravity-cast ingot

The tensile test results for the gravity-cast Al-Mg alloys are plotted against the Mg content in **Figure 8**. As the Mg content increased from 4.5–10%, the tensile stress gradually decreased from 231 to 185 MPa, and the 0.2% proof stress gradually increased from 106 to 138 MPa. The elongation decreased greatly from 18–10% when the Mg content increased from 4.5–6%, after which it linearly decreased with further increases to the Mg content, down to 3% at an Mg content of 10%. The Mg content had a greater effect on the elongation than on the tensile stress or the 0.2% proof stress.

3.3 Tensile test of hot-forged gravity-cast ingot

The results of the tensile test of the hot-forged gravity-cast ingot are plotted against the Mg content in **Figure 9**. The tensile stress, 0.2% proof stress, and

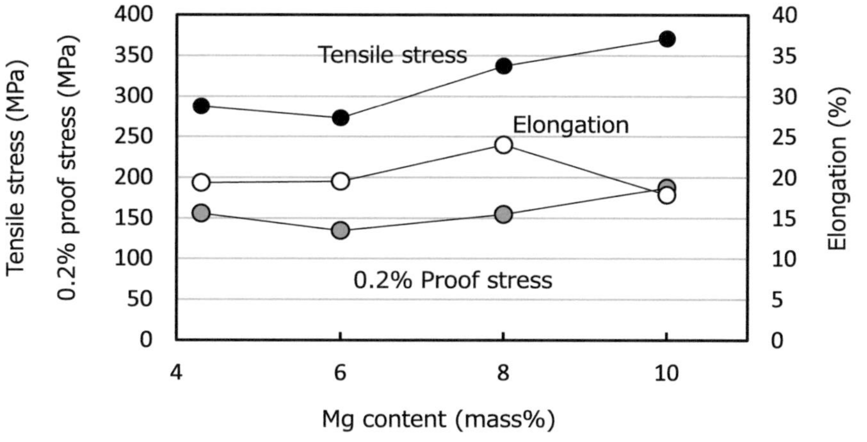

Figure 9.
Tensile test results for hot-forged gravity-cast Al-Mg alloys.

elongation of all Al-Mg alloys were increased by the hot forging. The tendencies of the tensile stress and elongation for the Mg content were also changed by the hot forging. The tensile stress increased with increasing Mg content. The increase (improvement) of the tensile stress became greater as the Mg content increased. At Mg contents of 4.5% and 10%, the tensile stress increased from 231 to 287 MPa and 185 to 270 MPa, respectively, which corresponds to respective increases of 56 and 185 MPa. When the Mg content was 8%, the elongation was maximized, and the elongation of the Al-10%Mg was the smallest among the alloys. The elongation of the hot-forged Al-8%Mg was 24%, and that of the Al-10%Mg was 17%. These results show that the hot-forged Al-Mg alloys have excellent strength and ductility.

Optical microscope images of the gravity-cast and hot-forged Al-Mg alloys are shown in **Figures 10** and **11**, respectively. The gravity-cast Al-Mg alloys had a dendrite microstructure, as shown in **Figure 10**; this is a typical structure for this type of casting. In contrast, there was not a dendrite structure in the hot-forged Al-Mg alloys, as shown in **Figure 11**, and the microstructure changed to a deformation structure as a result of the hot-forging. The grain size decreased as the Mg content increased until 8% Mg. This may contribute to the excellent mechanical properties of the Al-8%Mg alloy.

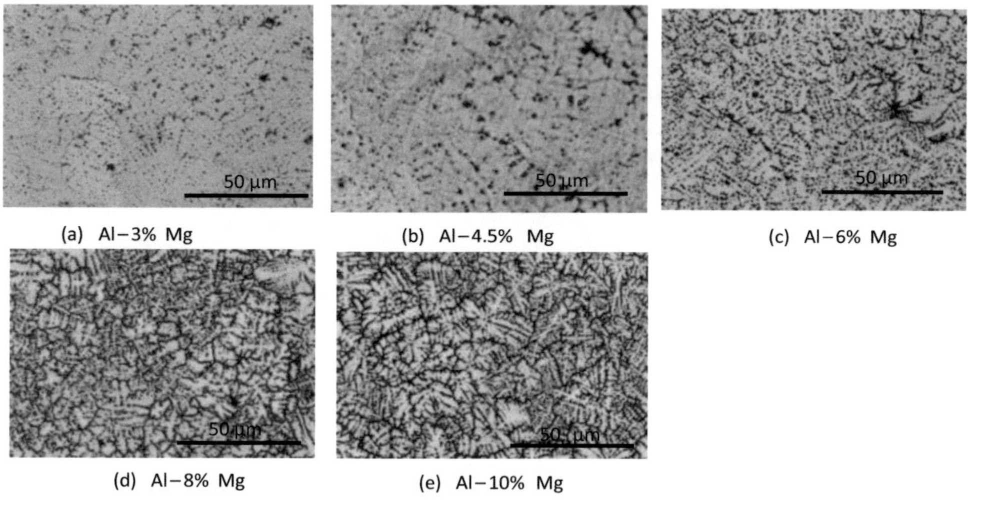

Figure 10.
Optical microscope images of gravity-cast Al-Mg alloys.

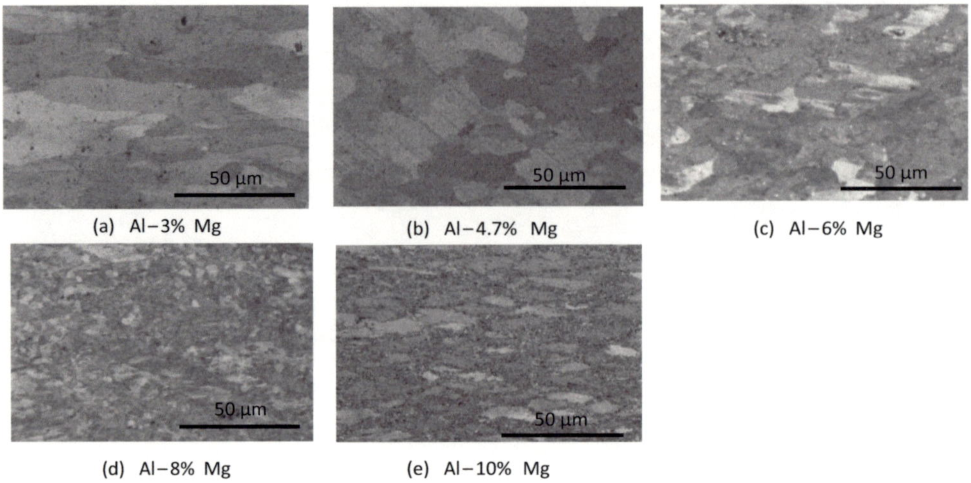

Figure 11.
Optical microscope images of hot-forged Al-Mg alloys.

3.4 Effect of Fe content on tensile test results for the hot-forged gravity-cast ingot

The tensile stress, 0.2% proof stress, and elongation of the different Al-Mg alloys are plotted against the added Fe content in **Figures 12–14**, respectively. The tensile stress was not influenced by the Fe content, as shown in **Figure 12**. The 0.2% proof stress of the Al-4.5%Mg was almost uniform, and that of other Al-Mg alloys increased gradually with increasing Fe content, as shown in **Figure 13**. The results are shown in **Figures 12** and **13** indicate that increasing the Fe content does not have a negative influence on the tensile stress or the 0.2% proof stress.

At Mg contents of 4.5%, 6%, and 10%, the elongation decreased with increasing Fe content, as shown in **Figure 14**. The elongations of the Al-8%Mg with 0.2% and 0.4%Fe were 27.6% and 24.6%, respectively, and the elongation of the Al-8%Mg without added Fe was 24.0%. When the Fe content was 0.2%,

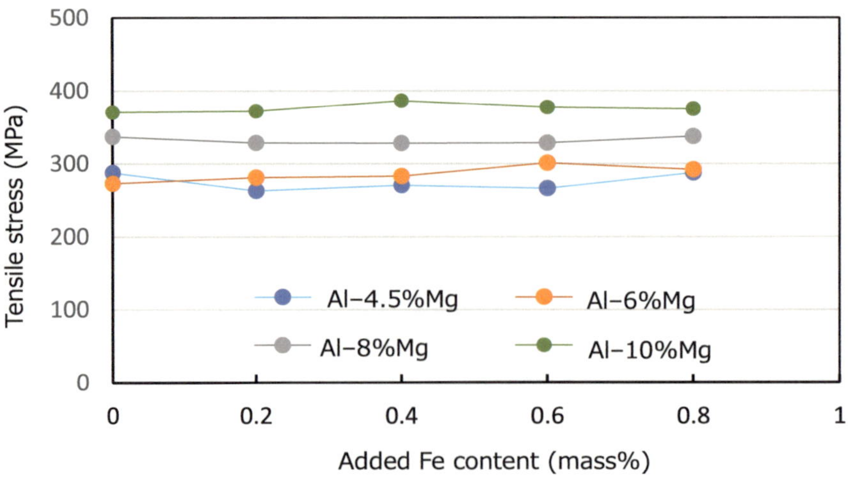

Figure 12.
Tensile stress of the hot-forged gravity-cast Al-Mg alloys plotted against Fe content.

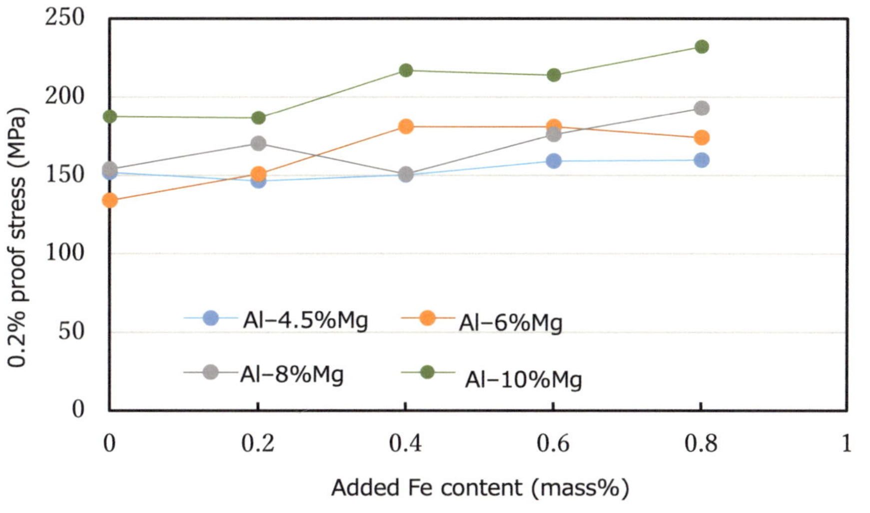

Figure 13.
The 0.2% proof stresses of the hot-forged gravity-cast Al-Mg alloys plotted against Fe content.

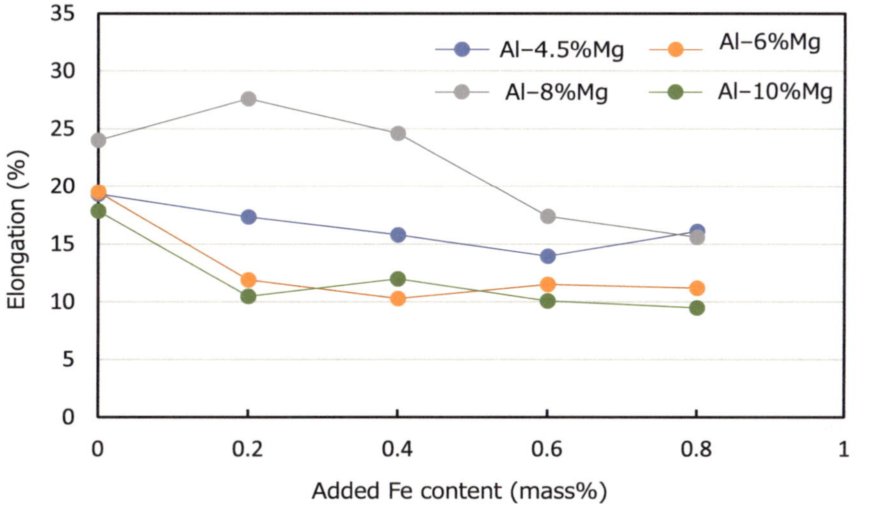

Figure 14.
Elongation of the hot-forged gravity-cast Al-Mg alloys plotted against Fe content.

the elongation did not decrease but increased. It is thought that the addition of Fe makes the elongation worse; however, in this case, the elongation increased. This means that when Fe impurities are incorporated during recycling, the elongation increases in comparison with that of the virgin alloy, demonstrating that upgrade recycling occurs. The elongation of the Al-8%Mg with 0.4%Fe was 24.6%, which means the added 0.4% Fe did not influence the elongation. The elongations of the Al-8%Mg with 0.6% and 0.8%Fe were 17.4% and 15.6%, respectively. The elongation of the hot-forged Al-8%Mg was excellent when the added Fe content was 0.8% or less. The elongation of the Al-10%Mg with 0.8%Fe was 9.5%, and that of the other Al-Mg alloys were greater than 9.5%. The hot-forged Al-Mg alloys have good elongation when Fe impurities are incorporated during recycling, with Al-8%Mg being particularly suitable for cast-forging and recycling.

4. Roll-casting

4.1 Single-roll caster equipped with a scraper

It is known that centerline segregation occurs between the solidification layers in Al-Mg alloy strips cast using a twin-roll caster (TRC). It is difficult to reproduce the occurrence of this type of centerline segregation in strips cast using twin-roll casters. In this study, a single-roll caster equipped with a scraper (SRCS) was used to cast Al-Mg alloys strips without centerline segregation [11].

In the SRCS, the molten metal is solidified on the side of the one roll, and a centerline does not form. The free solidified surface is scribed into a flat surface by the scraper. The scraper load was 0.2 N/mm, and no crack was formed on either surface of the strip because of the small scraper load. A copper roll was used to increase the cooling speed and roll speed. In the conventional TRC, steel rolls are used. The thermal conductivity of copper is much larger than that of steel, and the cooling ability of a copper roll is thus greater than that of a steel roll. The casting speed of the SRCS was 30 m/min, whereas the casting speed of a conventional TRC is usually slower than 2 m/min. The excellent cooling ability of the copper roll enabled high-speed roll casting. Schematic illustrations of the SRCS and the area near the scraper are shown in **Figure 15**. Cross-sections of Al-4.5%Mg strips cast using the high-speed TRC and the SRCS are shown in **Figure 16** [11, 12]. Centerline segregation occurred in the strip cast using the high-speed TRC and not in the strip cast using the SRCS.

4.2 Effect of Mg content on strip thickness and surfaces

Strips of Al-Mg alloys with Mg contents ranging from 4.5–10% could be continuously cast using the SRCS. The strip thickness is plotted against the Mg content in **Figure 17**. The strip became thicker as the Mg content increased. Two potential causes were considered to explain the relationship between the Mg content and the strip thickness. One is that the latent heat of the magnesium is smaller than that of the aluminum; thus, the latent heat of Al-Mg alloy decreases as the Mg content increases, which may then cause the strip thickness to increase with the Mg content. The other is that the amount of scribed and piled aluminum alloy under the scraper becomes greater as the Mg content increases, and the piled

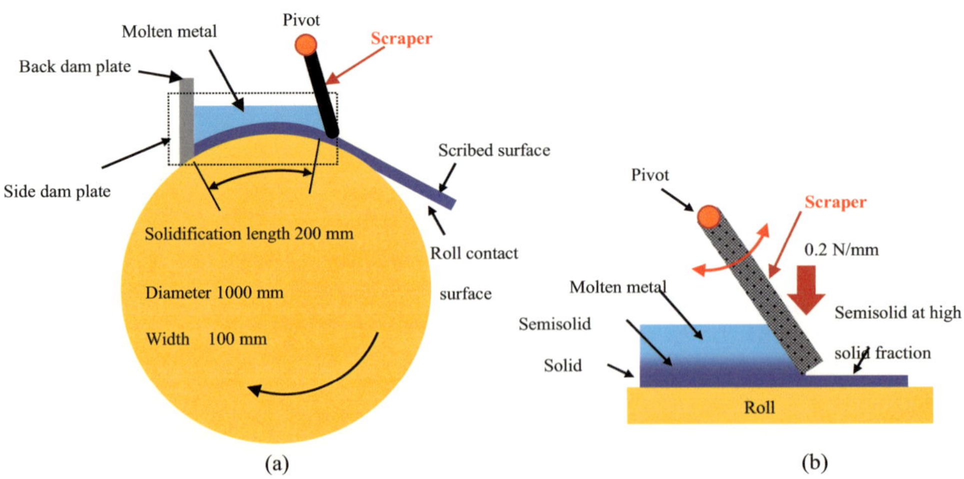

Figure 15.
Schematic illustrations of (a) a top-down view of a single-roll caster equipped with a scraper and (b) a view near the scraper.

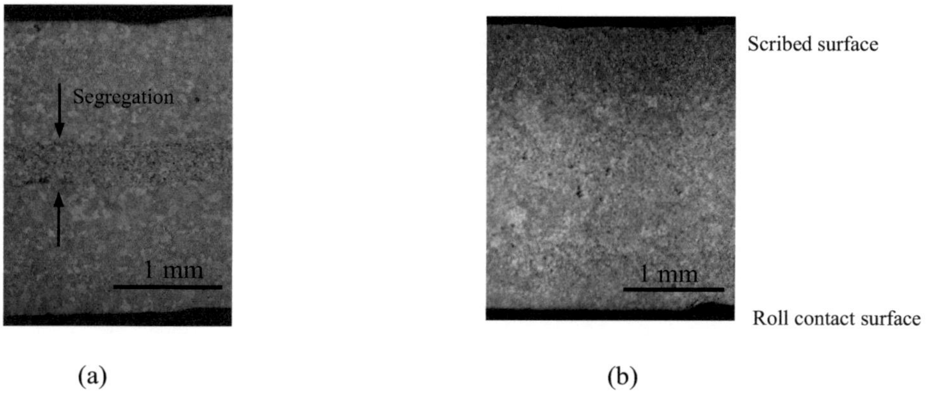

Figure 16.
Cross-sections of Al-4.5%Mg alloys cast using (a) a high-speed twin-roll caster and (b) a single-roll caster equipped with a scraper. The casting speed was 30 m/min.

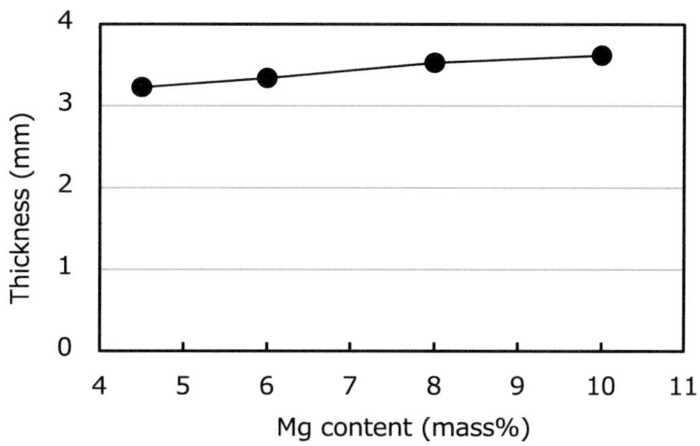

Figure 17.
Strip thickness plotted against Mg content.

aluminum alloy becomes a part of the strip [12]. Therefore, the strip becomes thicker as the Mg content increases.

The surfaces of the Al-Mg alloy strips are shown in **Figure 18**. The scribed surface did not have a metallic luster, whereas the roll contact surface did. The Mg content did not influence the surface condition.

4.3 Mechanical properties of roll cast Al-Mg alloy strips

The mechanical properties of the roll-cast Al-Mg alloy strips were tested by the tensile test. The cast strip was cold-rolled down to 1 mm and annealed at 360°C for 90 min. The dimensions of the test piece for the tensile test are shown in **Figure 19**.

The results of the tensile test are shown in **Figure 20**. The tensile stress increased monotonically at a gradual rate with increasing Mg content. The 0.2% proof stress was almost constant at different Mg contents. The elongation increased with increasing Mg content up to 8% Mg and then substantially decreased at 10% Mg. Comprehensively, judging from the tensile test, the Al-8%Mg showed the best mechanical properties.

A deep drawing test was then conducted to investigate the ability of sheet forming. The cast strip was cold-rolled down to 1 mm and annealed at 360°C for 90 min. The diameter of the punch used for the deep drawing test was 32 mm. The

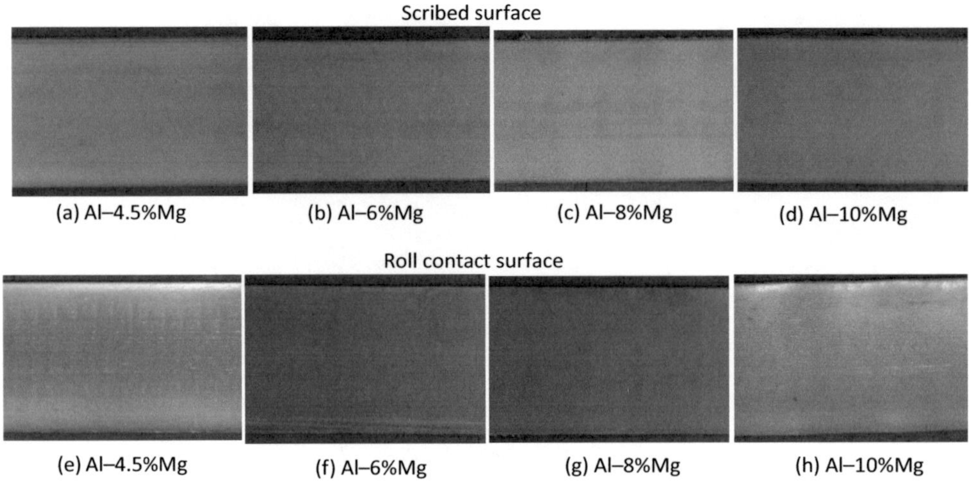

Figure 18.
Surfaces of as-cast Al-Mg alloy strips.

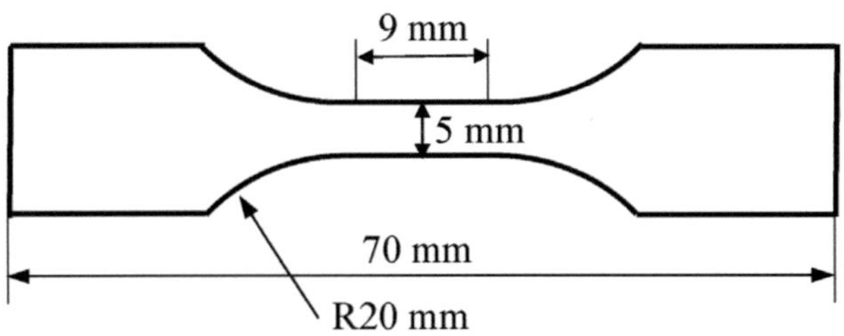

Figure 19.
Size of a test piece for the tensile test.

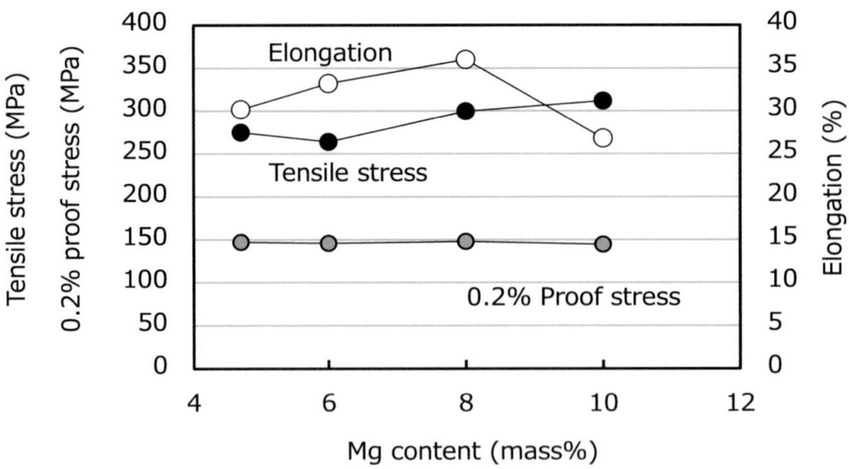

Figure 20.
Tensile test results for roll-cast Al-Mg alloys with different Mg contents.

deep drawing test was conducted under two conditions: with the roll-contact side of the strip facing outward and with the scribed surface facing outward. The results of the deep drawing test are shown in **Figure 21**. The limiting drawing ratio (LDR, the maximum ratio of circular blanks to the diameter of the die) of Al-4.5%Mg was 2.0 regardless of which side of the strip was facing outward. The LDR decreased

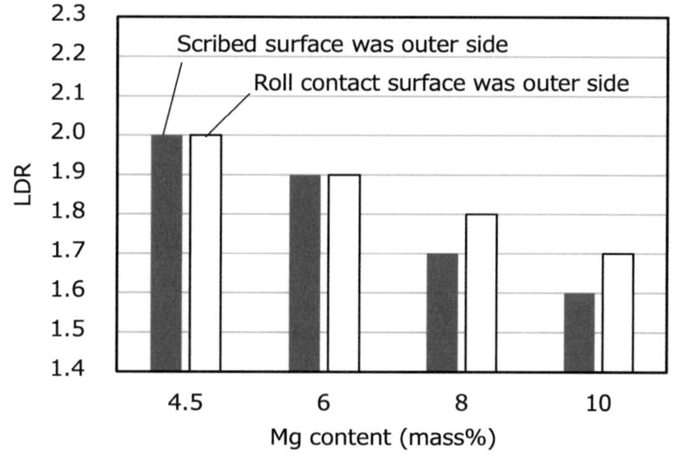

Figure 21.
Limiting drawing ratio at different Mg contents.

with increasing Mg content. When the Mg content was 4.5% or 6%, the LDR was not affected by which side faced outward; in contrast, when the Mg content was 8% or 10%, the LDR was better when the roll-contact side faced outward. The difference between the LDRs in these two cases is not suitable for sheet forming. The Al-4.5%Mg was most suitable for sheet forming. The optimal Mg content for deep drawing was different from that obtained from the elongation in the tensile test. These results demonstrate that 514.0 aluminum alloy is suitable for sheet forming, and 518.0 aluminum alloy is suitable for the easy shape plate, which needs strength and elongation. This shows that the choice of Mg content depends on the purpose. The forming ability is the most important property for sheets used in automobile manufacture, and thus the Al-4.5%Mg is suitable for this purpose. The Al-4.5%Mg was used to make the model alloy of recycled Al-Mg alloys.

4.4 Mechanical properties of Al-4.5%Mg with Fe

Impurities of 0.2%, 0.4%, 0.6%, and 0.8% Fe were added to the Al-4.5%Mg to model the recycled Al-Mg alloy. The Al-4.5%Mg with Fe could be cast into a strip, as the addition of the Fe did not affect the ability of the roll casting; however, the addition of the Fe makes the strip hard and brittle. Edge cracks with lengths of 3 mm or less occurred in the Al-4.5%Mg with 0.8%Fe, and the cold rolling could be conducted on the strip down to 1 mm without breaking. When the added Fe content was less than 0.6%, edge cracking did not occur. The surfaces of the as-cast and the cold-rolled strips of the Al-4.5%Mg and the Al-4.5%Mg with 0.8%Fe are shown in **Figure 22**. There was no difference between the scribed and roll-contact surfaces of the cold-rolled virgin Al-4.5%Mg and Al-4.5%Mg with 0.8%Fe strips. It is thought that the increase in the Fe content does not affect the surface properties of the Al-4.5%Mg sheet cast by the SRCS after cold rolling.

Cross-sections of the virgin Al-4.5%Mg and Al-4.5%Mg with 0.8%Fe strips are shown in **Figure 23**. The grain of the as-cast Al-4.5%Mg strip was almost uniform in the thickness direction. In the as-cast strip of Al-4.5%Mg with 0.8%Fe, the grain of the roll-contact side of the strip was finer than that of the scribed side. The effect of cooling speed on the grain size of the Al-4.5%Mg with 0.8%Fe was more apparent than in the Al-4.5%Mg. This is the influence of the added Fe. The Fe formed a crystal nucleus, and many crystals were made. As a result, the grain number increased and the grain size became small near the roll-contact side. The structure became a fine deformation structure after cold rolling and annealing.

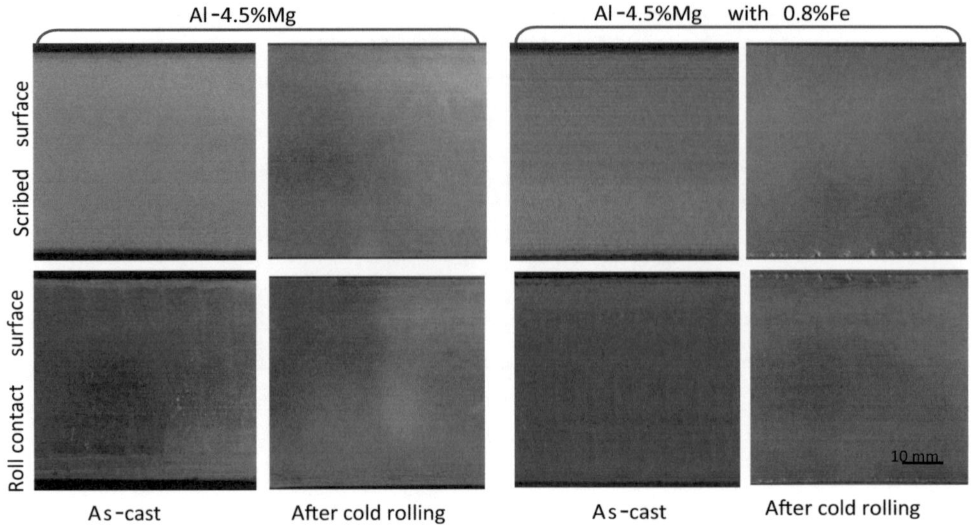

Figure 22.
Surfaces of as-cast and cold rolled strips of Al-4.5%Mg and Al-4.5%Mg with 8%Fe.

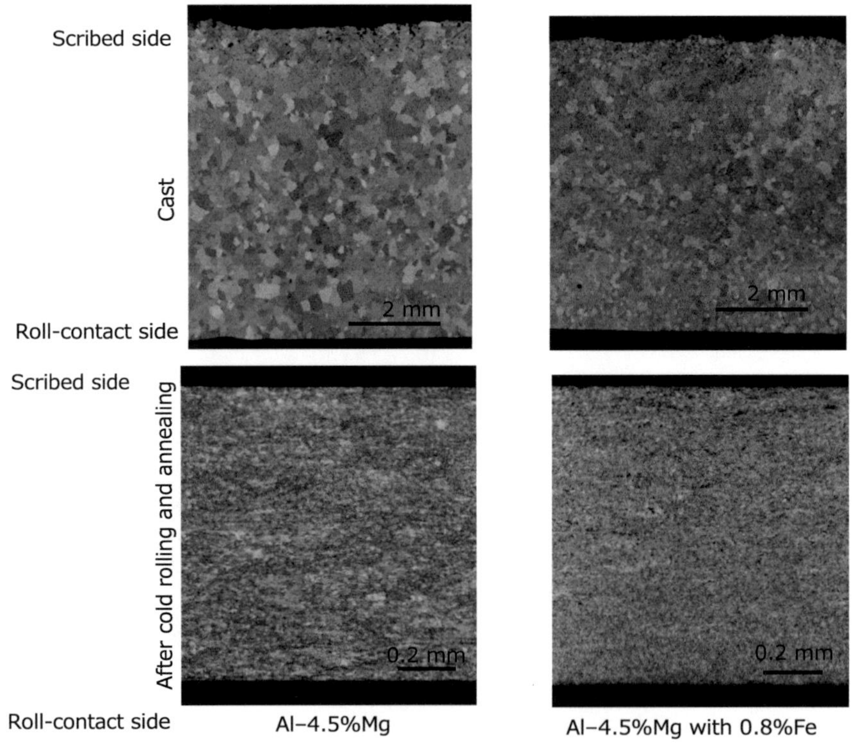

Figure 23.
Cross-sections of as-cast strip and cold rolled and annealed strip of Al-4.5%Mg and Al-4.5%Mg with 0.8%Fe. Annealing: 360°C for 90 min.

The results of the tensile test of the Al-4.5%Mg with Fe are shown in **Figure 24**. The tensile stress was almost uniform for the added Fe content. The 0.2% proof stress gradually increased with increasing Mg content, and the elongation very gradually decreased. The elongations of the Al-4.5%Mg and Al-4.5%Mg with 0.8%Fe were 30.3% and 28.6%, respectively. The reduction of the elongation with the addition of Fe

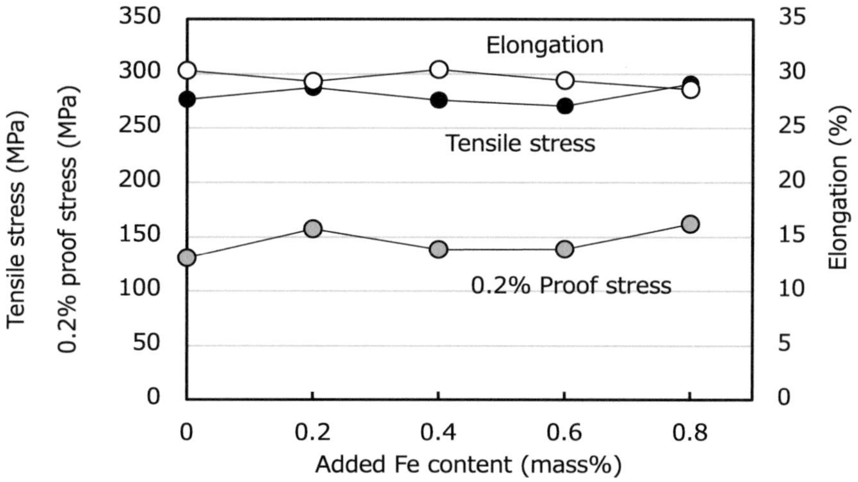

Figure 24.
Result of tensile test of Al-4.5%Mg with Fe.

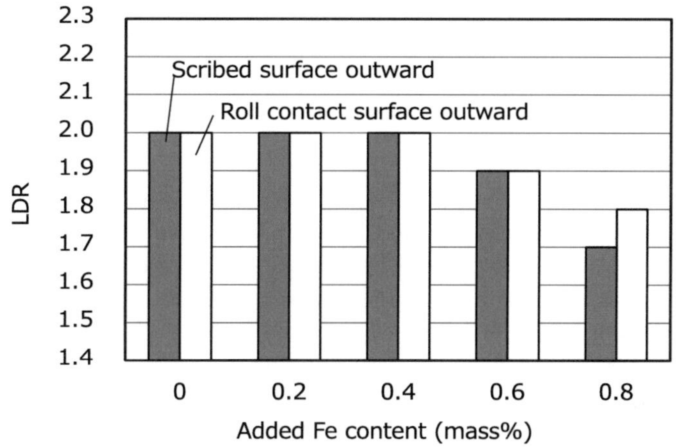

Figure 25.
Limiting drawing ratio of Al-4.5%Mg with different Fe Contents.

was very small. The intermetallic compound including Fe may be very fine because of the rapid solidification of the rolling caster, and it did not make the elongation worse.

The LDR of Al-4.5%Mg with Fe is shown in **Figure 25**. The LDR did not decrease from 2.0 until the addition of 0.4% Fe. When the Fe content was 0.6%, the LDR was 1.9. The LDR when the scribed surface faced outward was the same as that when the roll contact surface faced outward until the Fe content was 0.6%. Therefore, the ultimate addition of Fe to the Al-4.5%Mg was 0.4%.

5. Conclusions

5.1 Die casting

The fluidity of the Al-Mg alloy decreased with increasing Mg content until 6% Mg and then increased as the Mg content was increased beyond 6%. The fluidity was greatest at an Mg content of 10% and progressively decreased at contents of 4.5%, 8%, and 6%. The fluidity increased with increasing Fe content.

The tensile strength and the 0.2% proof stress increased with increasing Mg content. The elongation of the Al-6%Mg was greater than that of the Al-4.5%Mg. The addition of Fe did not degrade the tensile stress and the 0.2% proof stress. The elongation was reduced by the addition of Fe. The elongations of the Al-4.5%Mg with 0.8%Fe and Al-6%Mg with 0.8%Fe were 10% and 8%, respectively. These elongations were as good as those obtained during die casting. The amount of decrease in the elongation of the Al-4.5%Mg was smaller than that of the Al-6%Mg. The Al-4.5%Mg is suitable for recycling when the recycled alloy was die-cast.

5.2 Cast-forging

The mechanical properties of the gravity-cast Al-Mg alloys were increased by hot-forging. The tensile strength and the 0.2% proof stress increased as the Mg content increased. The elongation of Al-8%Mg was greater than that of the other Al-Mg alloys considered in this study. The addition of Fe did not degrade the tensile stress or the 0.2% proof stress. The elongation of all alloys except for the Al-8%Mg was degraded by the addition of the Fe. The elongation of the Al-8%Mg with 0.2%Fe was greater than that of the Al-8%Mg, and that of the Al-8%Mg with 0.4%Fe was the same as that of the Al-8%Mg. The elongation of the Al-8%Mg with 0.8%Fe was 15.6%, which is sufficiently large for a forged aluminum alloy. Al-8%Mg is thus suitable for cast-forging.

5.3 Roll casting

As the Mg content was increased, the tensile stress gradually increased, whereas the 0.2% proof stress remained almost constant. The elongation of the Al-8%Mg was greater than those of the other Al-Mg alloys. The LDR decreased as the Mg content increased. Therefore, the Al-4.5%Mg was selected as the most suitable for sheet forming among the roll-cast Al-Mg alloys. The tensile stress of the Al-4.5%Mg was almost uniform for the added Fe content. The 0.2% proof stress gradually increased with increasing Mg content, and the elongation very gradually decreased. The elongations of the Al-4.5%Mg and Al-4.5%Mg with 0.8%Fe were 30.3% and 28.6%, respectively. The LDR was 2.0 until an Fe content of 0.4%, and then it decreased with increasing Fe content for Fe contents greater than 0.6%. This shows that twice-recycled Al-4.5%Mg (Al-4.5%Mg with 0.4%Fe) cast using the roll-caster can be used for sheet forming.

Acknowledgements

This work was supported by SUZUKI FOUNDATION.

Author details

Toshio Haga
Department of Mechanical Engineering, Osaka Institute of Technology, Osaka, Japan

*Address all correspondence to: toshio.haga@oit.ac.jp

IntechOpen

© 2021 The Author(s). Licensee IntechOpen. This chapter is distributed under the terms of the Creative Commons Attribution License (http://creativecommons.org/licenses/by/3.0), which permits unrestricted use, distribution, and reproduction in any medium, provided the original work is properly cited. CC BY

References

[1] Haga T, Ikawa M, Kumai S. A high speed twin roll caster for aluminum alloy thin strip. MRS-J. 2004;**29**: 1823-1828

[2] Haga T, Ikawa M, Kumai S, Watari H. Formability of roll cast recycled aluminum alloy. Key Engineering Materials. 2007;**340-341**:605-610. DOI: 10.4028/www.scientific.net/KEM.340-341.605

[3] Kumar S, Hari Babu N, Scamans MS, Fan Z, O'Reilly KAQ. Twin roll casting of Al-Mg alloy with high added impurity content. Metallurgical and Materials Transactions A: Physical Metallurgy and Materials Science. 2014;**45**:2842-2854. DOI: 10.1007/s11661-014-2229-x

[4] Mansurov YN, Rakhmonov JU, Letyagin NV, Finogeyev AS. Influence of impurity elements on the casting properties of Al-Mg based alloys. Non-Ferrous Metals. 2018;**1**:24-29

[5] Zhu X, Blake P, Dou K, Ji S. Strengthening die-cast Al-Mg and Al-Mg-Mn alloys with Fe as a beneficial element. Materials Science and Engineering A. 2018;732:240-250. DOI: 10.1016/j.msea.2018.07.005

[6] Phongphisutthinan C, Tezuka H, Sto T. Semi-solid microstructure control of wrought Al-Mg-Si based alloys with Fe and Mn additions in deformation-semi-solid-forming process. Materials Transactions. 2011;**52**:834-841. DOI: 10.2320/matertrans.L-MZ201119

[7] Birol Y, Akdi S. Cooling slope casting to produce EN AW 6082 forging stock for manufacture of suspension components. Transactions of Nonferrous Metals Society of China. 2014;**5**:1674-1682. DOI: 10.1016/S1003-6326(14)63240-4

[8] Ono H, Tomita S, Ishii H, Ogasawara A. Forging shape in application of cast/forging process to Sl-Si-Mg alloy suspension parts. Journal of Japan Institute of Light Metals. 1995;**45**:187-192

[9] Chang YL, Hung FY, Lui TS. Study of microstructure and tensile properties of infrared-heat treated cast-forged 6082 aluminum alloy. Journal of Materials Research. 2019;**8**:173-179. DOI: 10.1016/j.jmrt.2017.10.004

[10] Zhang Q, Cao M, Zhang D, Zhang S, Sun J. Research on integrated casting and forging process of aluminum automobile wheel. Advances in Mechanical Engineering. 2014. DOI: 10.1155/2014/870182

[11] Haga T, Tsukuda K, Oida K, Watari H, Nishida S. Casting of Al-Mg strip using single roll caster equipped with a scraper. Key Engineering Materials. 2021;**880**:49-56. DOI: 10.4028/www.scientific.net/KEM.880.49

[12] Haga T. High speed roll caster for aluminum alloy. Metals. 2021;**11**:520. DOI: 10.3390/met11030520

Chapter 4

Characterization of Casting Properties of Rare-Earth Modified A356

Kerem Can Dizdar, Hayati Sahin, Furkan Tezer and Derya Dispinar

Abstract

Al-Si alloys are an attractive choice of light alloys due to their low density and high mechanical properties. The application areas include automotive and aerospace parts. With the increased requirements and demands on such applications, the researchers have focused on the enhancement of properties with the addition of novel alloying elements. In the last decade, modification of microstructure by rare-earth element additions has become popular. In this work, systematical tests were carried out by using fluidity test mold that has 8 arms with different thicknesses and tensile test mold that contains 10 bars to statistically analyze the mechanical properties. Different ratios ranging from 0.05 to 1 wt% of Er, Sc, Y, and V were added to Al-7Si alloys. The addition of 0.1 wt% Sc shows the highest fluidity whereas V additions have the lowest. Statistically, Sc addition revealed the highest reproducible results in terms of tensile properties where Y had the highest scatter. Er additions have the highest UTS and elongation at fracture.

Keywords: casting, A357, rare earth, tensile, fluidity, microstructure

1. Introduction

Grain refinement is a method used to alter the microstructure from heterogeneous dendrites into smaller and globular morphology. In this way, the casting properties are enhanced significantly. The feedability of the melt increases and thus fluidity increases. Since the interdendritic flow of the liquid resistance is decreased, the shrinkage and porosity decrease. Additionally, obtaining homogeneous and refined grains lead to achieving higher mechanical properties. It is important to note that the reduction of casting defects has the most determining influence on mechanical properties as well as the homogeneous distribution of secondary phases along with the microstructure.

Over the past several years, the modification of aluminum alloys with rare earth elements has gained great attention. The studies are concentrated on the improvement of mechanical properties by grain refinement or modification of Si morphology, particularly in Al-Si alloys. The use of Sc as a grain refiner has almost proven itself to be a key grain refinement agent. However, the expense of using Sc and the high cost have led the application to be limited. Alternative elements such as V, Y,

Er, Eu and Ce have been started to be investigated in detail. Their refinement and modification effects have been found in Al–Si–Mg alloys [1–6].

Mazahery [7] investigated the effect of 600 ppm Eu addition to A360 on mechanical properties. It was found that as SDAS was increased, UTS was decreased by 25% and elongation at fracture was decreased by half. The microstructure and mechanical property change of A356 by Eu additions between 200 and 1000 ppm were investigated by F. Mao [8]. Eu was an effective Si modifier. The eutectic temperature was decreased, and Si morphology was altered from coarse plates into smaller fine spherical fibers which increased the tensile properties by 15% when 1000 ppm Eu was used.

L. Li [9] compared the semisolid die casting method to gravity casting. It was reported that 0.1 wt% RE addition had not affected the grain size but modification of Si to fibrous morphology had resulted in increased elongation at fracture for both methods. Increasing RE content to 0.4 wt% had resulted in the formation of needle-like intermetallic and the material had become brittle. Up to 0.8 wt%, Yb addition had formed globular primary and secondary dendrites with both Si and Fe intermetallic converted from acicular to finer morphology, however, higher amounts of addition of Yb to Al12Si had resulted in decreased mechanical properties [10]. Kabliman [11] added Sc to Al-Mg alloys to produce alternative foil material to Al-Cu alloys where 550 MPa was achieved at 0.5 mm thickness. Elgallad [12] found that in the absence of Sr., the addition of La and Ce to A356 and A413 had not modified Si, instead primary Si phases were formed. In the presence of Sr., much finer Si modification was observed. On the other hand, when Ti was added to these alloys, a high population of intermetallic formation was observed. When Yb was added together with La to A356, spherical Si particles were formed with a depression of 8°C of eutectic temperature [13]. M. Colombo [14] added different amounts of Er to A356 and showed that Er can both grain refine and modify Si at the same time. Best results were obtained at 0.3 wt% Er addition, and increased Er addition had resulted in the formation of intermetallic phases which decreased the ductility of A356 [15–17].

The effect of RE additions on the porosity formation of A356 alloy was investigated by Elgallad [18]. Various amounts of La, Ce, and La + Ce modifications were tried. When the intermetallic was increased (3%), the feedability was decreased which increased the shrinkage and porosity formation. On the other hand, it was reported that Ce was more dominant in pore formation than La. It was also found that Sr. was still the most effective modifier, but it also had a high affinity to RE elements possibly resulting in intermetallic formation [1]. Z. Wang [19] found that the addition of 0.1 wt% La and Ce increased the ductility of the alloy. The formation of fine and distributed secondary phases was observed on the fracture surface as dimples however in the absence of the RE added alloy, the fracture surface consisted of brittle cracks. Asmael [20] investigated the microstructure and mechanical properties of Ce added Al11SiCuMg alloy. Tensile strength and quality index were found to be the highest at 0.1 wt% Ce modified alloy. SDAS was decreased by 36%. When Ce content was changed between 0.5 and 1.0 wt%, it was found that there was no change in the microstructure and tensile properties. Pourbahari [21] studied the properties of thin-walled castings of A357 with La addition. La content was varied between 100 and 1000 ppm. The grain size was reported to be decreased by 50% with an increase of approximately 80% UTS. Sc additions up to 0.4 wt% revealed 50% decrease in SDAS of A356 with obtaining of spherical Si morphology [22]. Thus, the eutectic phase was increased from 40 to 45% with increased tensile property from 180 to 300 MPa. Yii [23] added Ce to Al20Si hypereutetic alloy and reported that wear properties were increased when 1.6 wt% Ce was added. Tzeng [24] added Sc to Al11Si and reported that beta-Fe intermetallic was converted to Chinese script which did not reduce the ductility of the alloy. The 0.8 Sc refined microstructure eliminated Fe-intermetallic formation increased tensile properties [25].

In this study, different amounts of Er, Sc, Y, and V were added to the A356 alloy. Efforts were made to statistically compare the effect of RE additions on the mechanical properties of A356 alloy. The reproducibility of tensile properties was investigated. The change in the fluidity of alloys was also investigated in the octopus design [26, 27].

2. Experimental

The chemical composition of the A357 aluminum casting alloy studied within the scope of the study is given in **Table 1** (Quantolux QLX3). Ten cylindrical bars were produced and cast into a sand mold where the dimension of bars was Ø8 by 160 mm (**Figure 1**). The weight of the charge to be melted was measured (approximately 15 kg) and melted in an ICS induction furnace in an A50 SiC crucible at 750°C. Reduced

Si	Fe	Cu	Mn	Mg	Ti	Al
7.485	0.173	0.021	0.018	0.529	0.126	91.45

Table 1.
Optical emission spectrometer analysis.

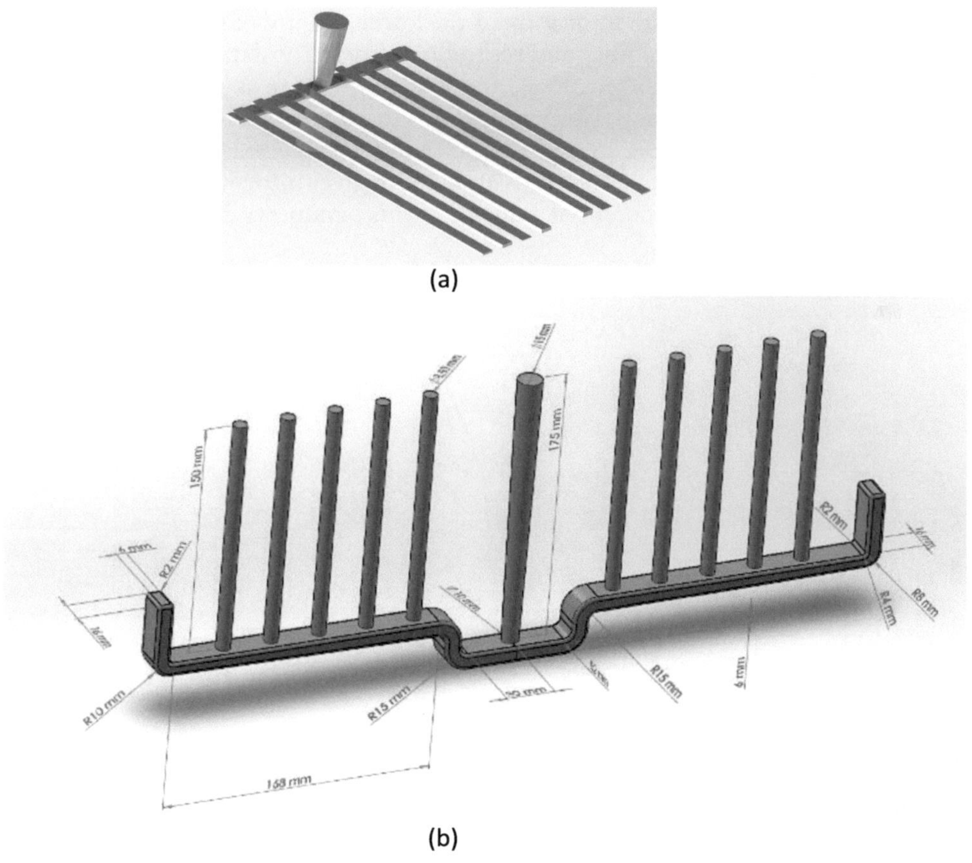

Figure 1.
Molds used for the characterization tests (a) fluidity (b) tensile bars.

pressure test samples were collected for measuring the melt cleanliness level. Ceramic diffusor lance was used where nitrogen gas was purged through the melt for the cleaning process.

The master alloys added to the melt were Al3Er, Al2Sc, Al5V and Al10Y. The weight % was aimed to achieve 0.05 and 0.1 in the alloy. Sand molds were prepared for the characterization tests. Octopus mold design [27] (**Figure 1a**) was used for fluidity characteristic where each arm had a length of 500 mm with thickness varying from 0.2 to 8 mm. Ten tensile bars were produced for each parameter where 10 mm diameter and 180 mm length cylinders were produced (**Figure 1b**). The molds were placed next to each other and after tensile bars were cast, the fluidity test mold was filled. This was repeated twice, and the same procedure was followed for each parameter to be able to have the same condition for each casting trial.

After the castings were complete, the samples were subjected to a heat treatment procedure [28] where the solutionizing was carried out at 540°C for 6 hours. The samples were then quenched in water (80°C) followed by 4 hours of aging at 145°C. Samples were machined according to ASTM E8/E8M standard and tensile tests were carried in Zwick Roell 8596. Weibull analysis was used to plot the survivability of the alloys studied in this work.

3. Results and Discussion

3.1 Tensile properties

The survivability plot is an indication of the lower and upper limits of any variable which gives the reliability and reproducibility of the data. In **Figures 2–4**, the tensile property changes of Sc, Er, Y, and V added A357 in as-cast conditions are given. For the yield strength (**Figure 2**), 0.05 and 0.1 wt% Sc reveal the highest reproducible results where the strength value changes between 116 and 120 MPa. In terms of design parameters, the survivability plot gives information about the possible lowest and highest potential values that this parameter can provide. For

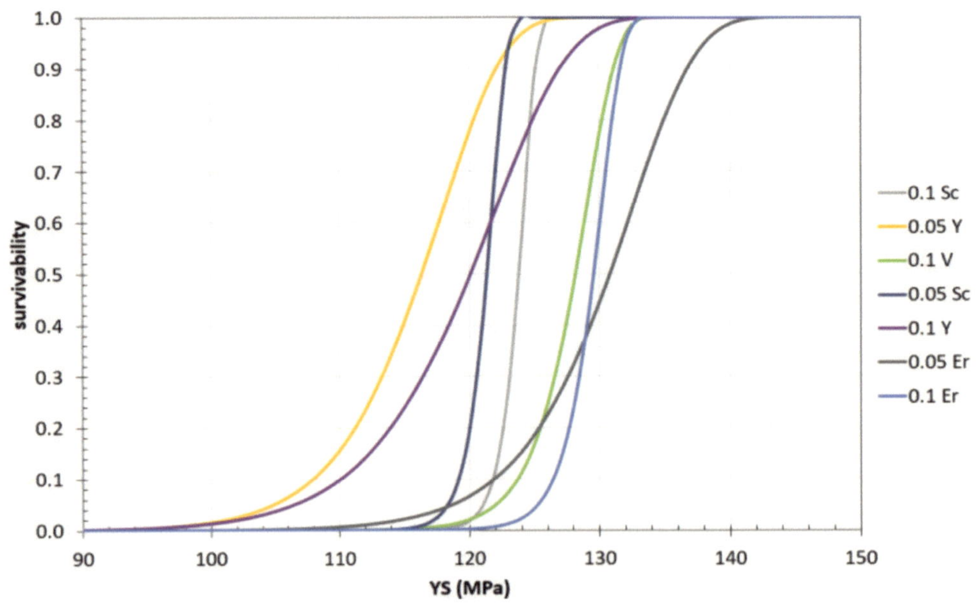

Figure 2.
Survivability plot of yield strength.

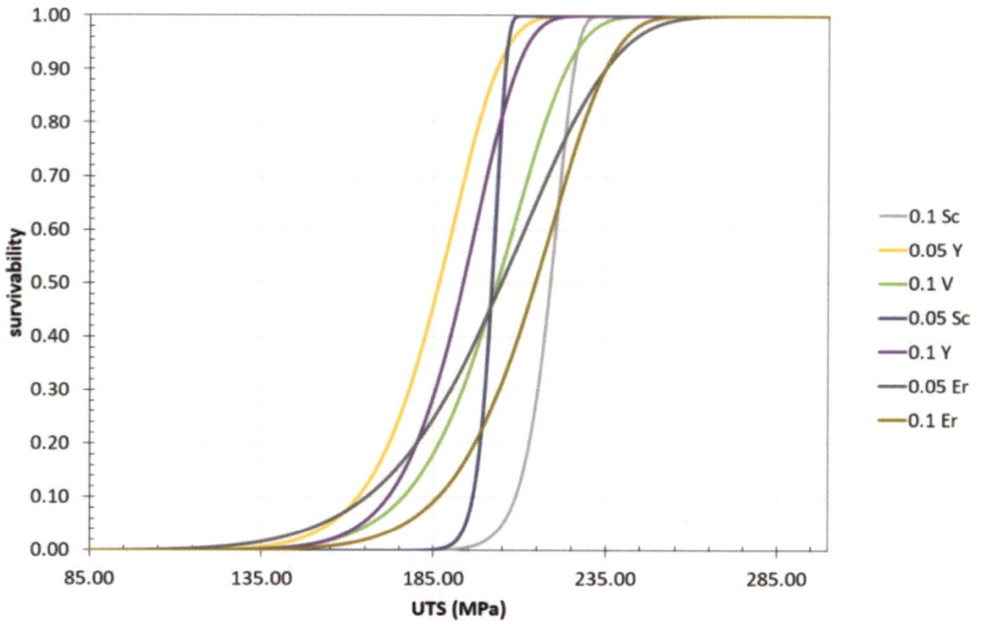

Figure 3.
Survivability plot of ultimate tensile strength.

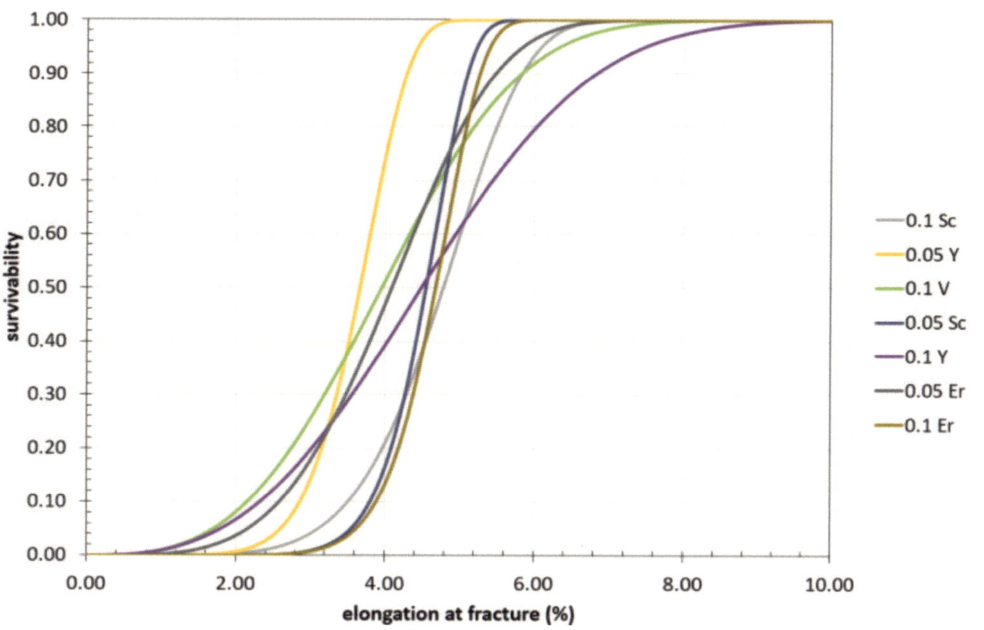

Figure 4.
Survivability plot of elongation at fracture.

example, 0.05 wt% Er castings reveal that they can exhibit a minimum of 109 MPa below which the material will always be in an elastic region. This particular test result also provides the information that this alloy can provide a yield strength of 140 MPa as well. Thus, it shows the potential of the alloy. However, since the scatter is too high (yield strength ranging between 109 and 140 MPa), the reliability of this parameter becomes very low. On the other hand, looking at 0.05 wt% Sc addition, it is very steep at values between 116 and 120 MPa which is quite narrow and thus the reliability of this casting is too high. This states that the yield strength value will undoubtedly be between these values.

The highest scatter is observed with 0.05 wt% Er and 0.1 wt% Y additions. The values are in the range of 90–133 MPa for Y, and 107–140 MPa for Er. Although, Er shows high scatter, yet it had the potential that this addition has the highest yield strength value of 140 MPa. On the other hand, the lowest yield strength value was obtained when Y was used as a modifier to A357.

The ultimate tensile strength values for all additions were very close to each other (**Figure 3**). The highest reliable and most reproducible results were obtained with 0.05 wt% Sc samples which were 197 MPa with a standard deviation of ±3. Pramod [22] had found that Sc additions to A356 had increased the UTS 45% when Sc was added up to 0.4 wt%.

The highest scatter was observed 0.05 wt% Er castings. The lowest UTS values were found to be in Y modified alloy whereas the highest possible UTS values were obtained with 0.05 and 0.1 wt% Er added A357 with a potential of having 265 MPa. Colombo [14] reported that 0.3 wt% Er addition was the optimum ratio in their study to exhibit the highest tensile properties.

As can be seen in **Figure 4**, 0.1 wt% Y added A357 had the potential of having 9% elongation at fracture, however, this alloy has the highest scatter and thereby revealing the least reliable casting. The 0.05 wt% Y added A357 has the lowest elongation while 0.05 and 0.1 wt% Er show the highest reproducible results.

The change in the tensile properties after the samples were subjected to heat treatment is given in **Figures 4–9**. There is almost twice the change in yield strength after heat treatment with Sc showing the highest value of 260 MPa and Y showing the lowest of 216 MPa (**Figure 5**). The ultimate tensile strength increases approximately 25% where Sc and Er have the highest value of 300 MPa whereas Y additions show the lowest UTS of 240 MPa (**Figure 6**). There is an almost 70% decrease in elongation values for all additions, except Er additions. Therefore, the toughness change of Er with and with heat treatment is nearly too small and Er additions exhibit the highest toughness amongst the alloys studied in this work (**Figure 7**).

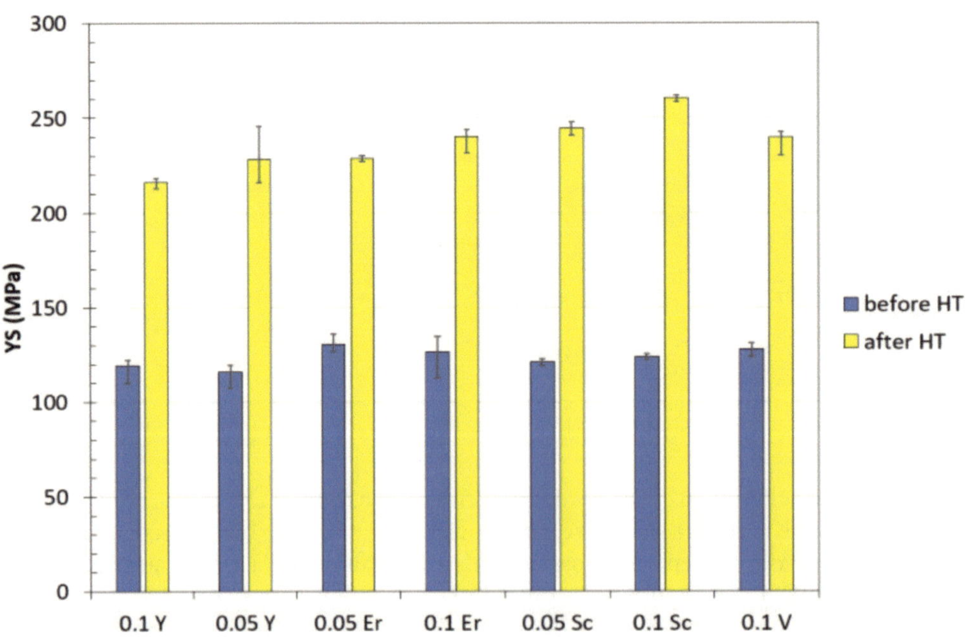

Figure 5.
Yield strength change with heat treatment.

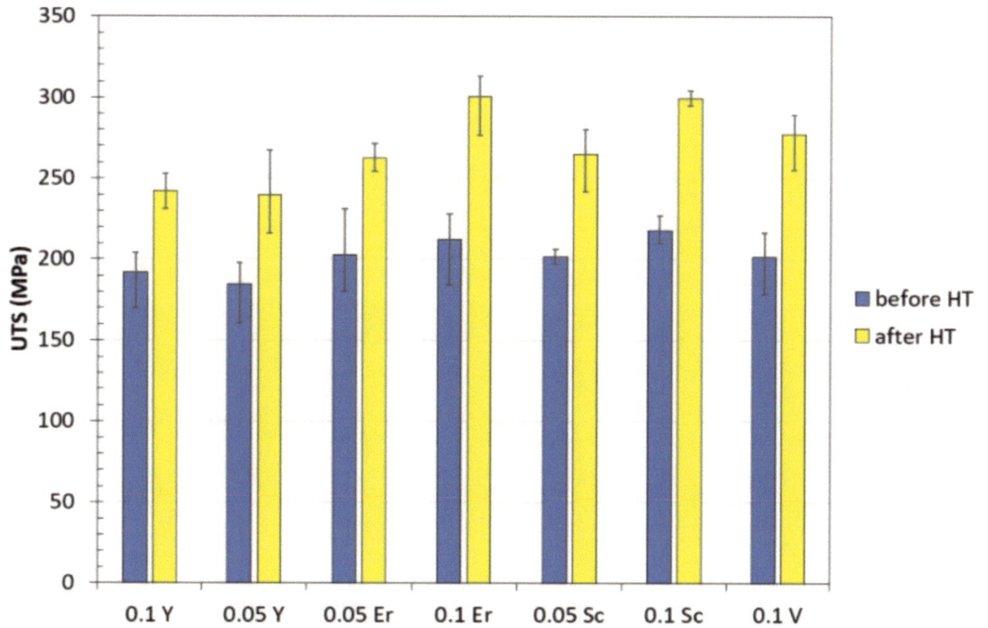

Figure 6.
Ultimate tensile strength change with heat treatment.

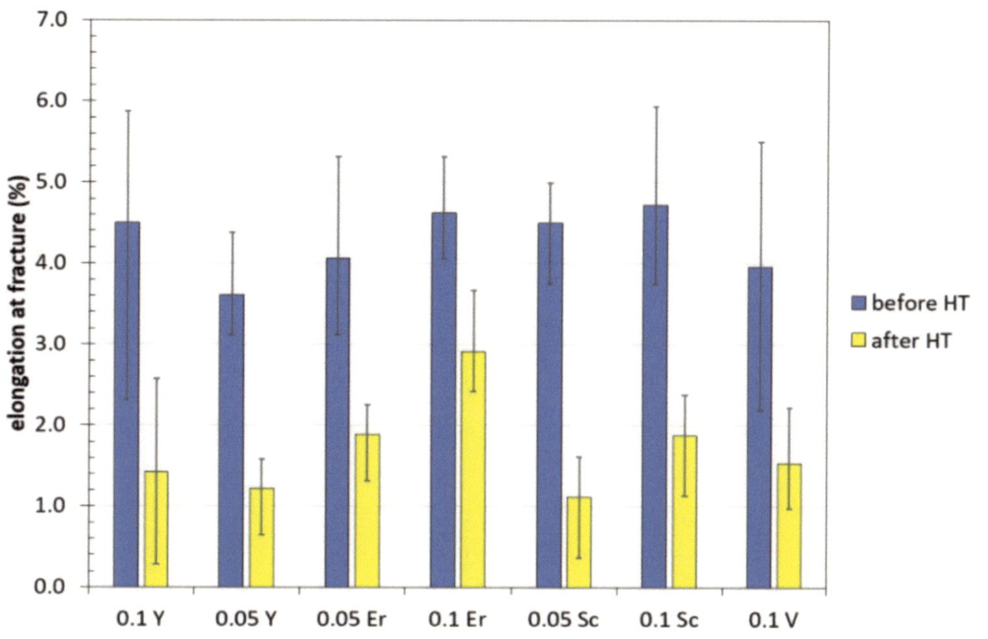

Figure 7.
Elongation at fracture change with heat treatment.

Hu [29] used the HPDC method to check the mechanical properties change of Al12Si alloy where Er content was changed between 0, 0.3, 0.06, and 0.9 wt%. It was reported that as Er content was increased, UTS was increased with decreased grain size, however, elongation at fracture values was decreased. This was attributed to the presence of Al_3Er phases found on the SEM images of fracture surface analysis. Therefore, it was concluded that these precipitates acted as reinforcements to the matrix, thus increasing UTS but decreasing elongation. Gao [16] reported

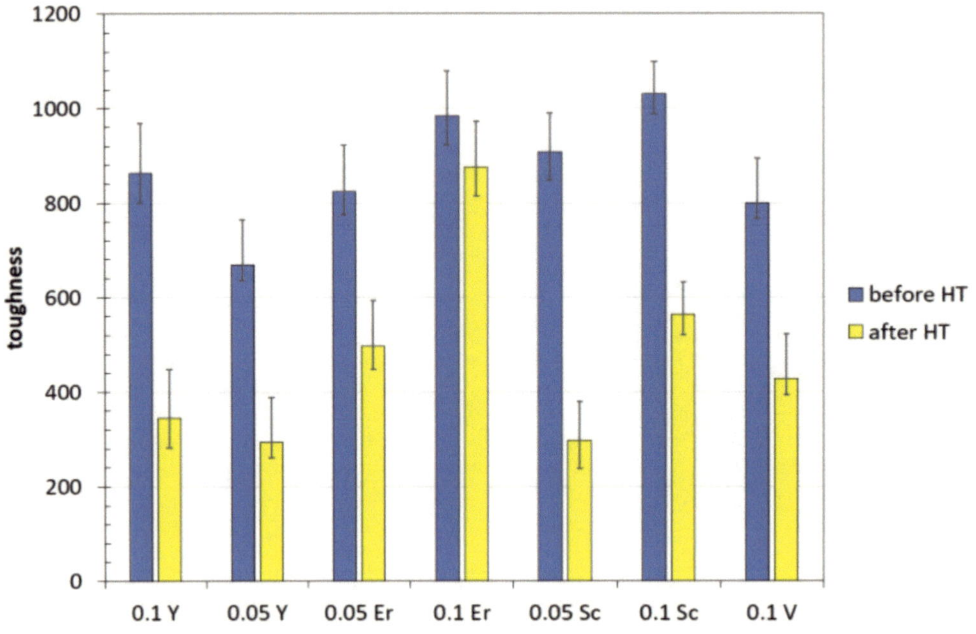

Figure 8.
Toughness change with heat treatment.

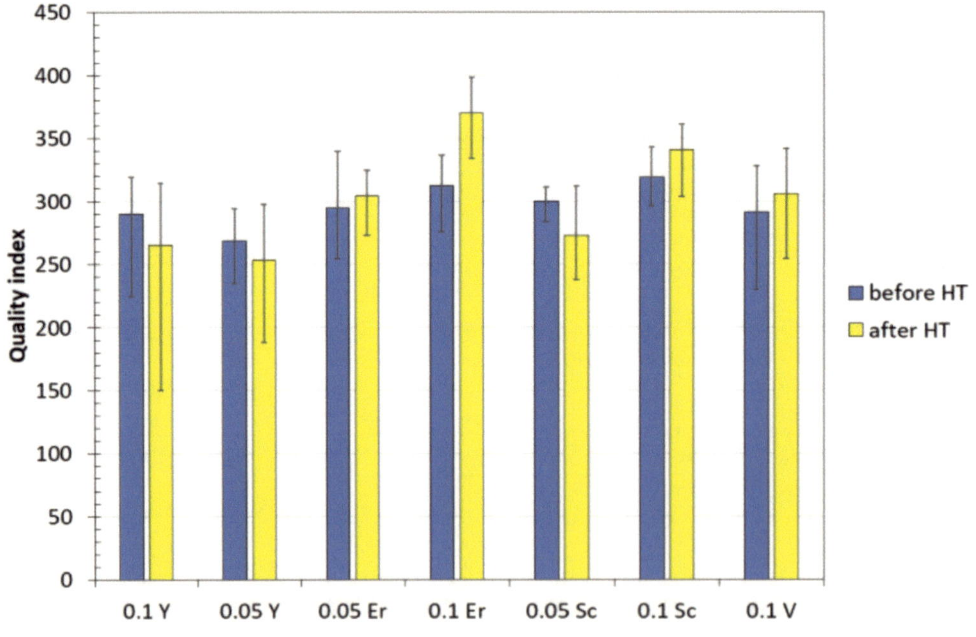

Figure 9.
Quality index change with heat treatment.

similar results claiming that nano-sized Al3Er, Al3Zr, and Al3(Zr,Er) precipitates show resistance to dislocations and increase the strength of pure aluminum significantly. Shi [15] proposed that Al3(RE) intermetallic phases do not have the grain refinement effect but rather act as barriers to dislocation movement and thereby increase the tensile properties of Al-Si alloys. Shi [15] also reported that Er has more physicochemical activity than that of other modifiers which have the potential to eliminate bifilms [27, 30–36] and increase the melt quality, therefore exhibiting higher mechanical properties. In the trials reported by Pramod [22], 0.4 wt% Sc addition to A356 had resulted in achieving 300 MPa levels with the lowest SDAS (10 μm) and refined Si morphology. In this work, 30 μm SDAS was good enough to achieve 300 MPa.

Pourbahari [21] studied the effect of La addition and they reported that the presence of intermetallic like Al_4La was more spherical and distributed in the microstructure (at low levels of La addition), and therefore they did not act as stress rising locations that would decrease the ductility of the alloy. However, when La content was higher than 0.1 wt%, new intermetallic phases such as AlSiLa were formed which were flakey and needle-like, thereby reducing the toughness of the matrix. The fracture surfaces were quasi-cleavage facets and not dimple like as in low La levels.

Drouzy's quality index [37] assessment was also used in this study and the results are given in **Figure 9**. It can be seen that Er and Sc both have the highest index values. Additionally, those elements are the only ones that show an increase in the quality index after heat treatment.

3.2 Microstructure

Although, the mechanical properties of each different grain refiner addition show a wide range of values and there is a quite difference between the samples, yet the microstructural analysis had revealed that the DAS and SDAS of all castings were very close to each other (**Figures 10** and **11**). Only 0.05 wt% Y addition shows the highest DAS value compare to the others, but all SDAS measurements of castings were recorded to be between 30 and 40 μm as seen in **Figure 12**. It is interesting to note that only in Y additions Si morphology was altered to fibrous type (**Figures 10b** and **11b**). Colombo [14] added 0.2 and 0.4 wt% Er and found that Er acted as a modifier for Si

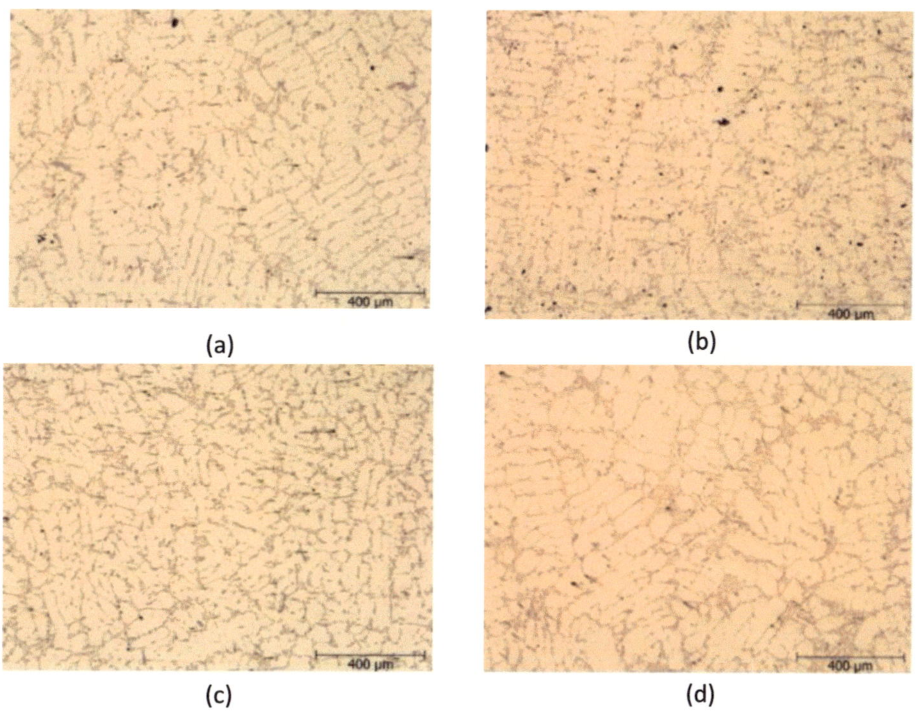

Figure 10.
Microstructures of 0.05 wt% (a) V, (b) Sc, (c) Er and (d) Y added A356.

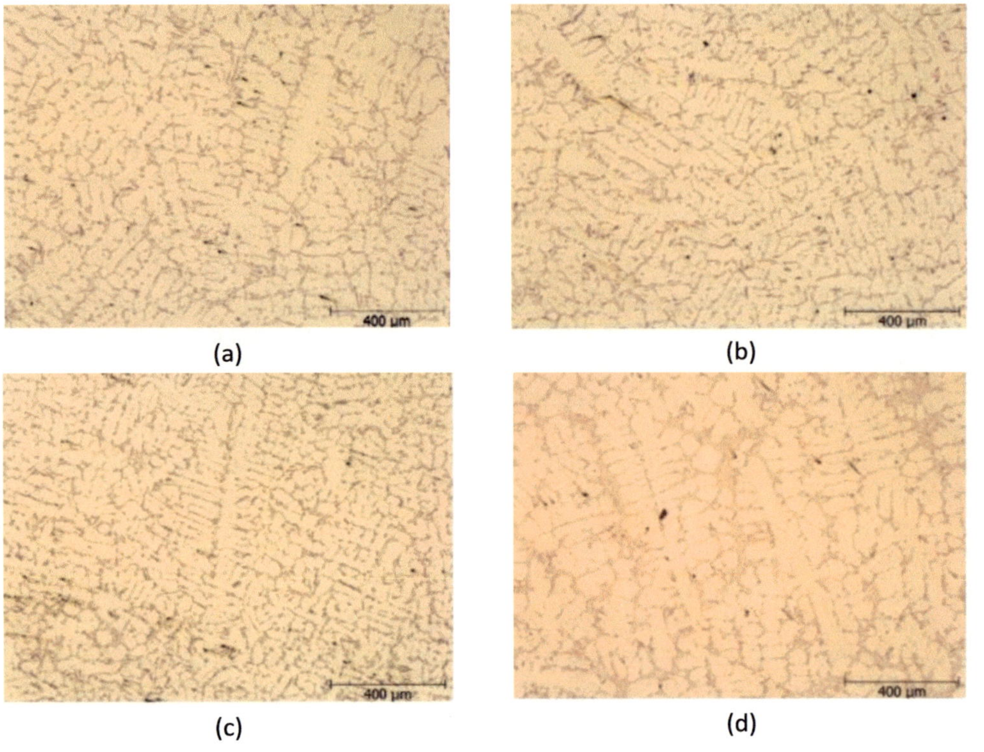

Figure 11.
Microstructures of 0.1 wt% (a) V, (b) Sc, (c) Er and (d) Y added A356.

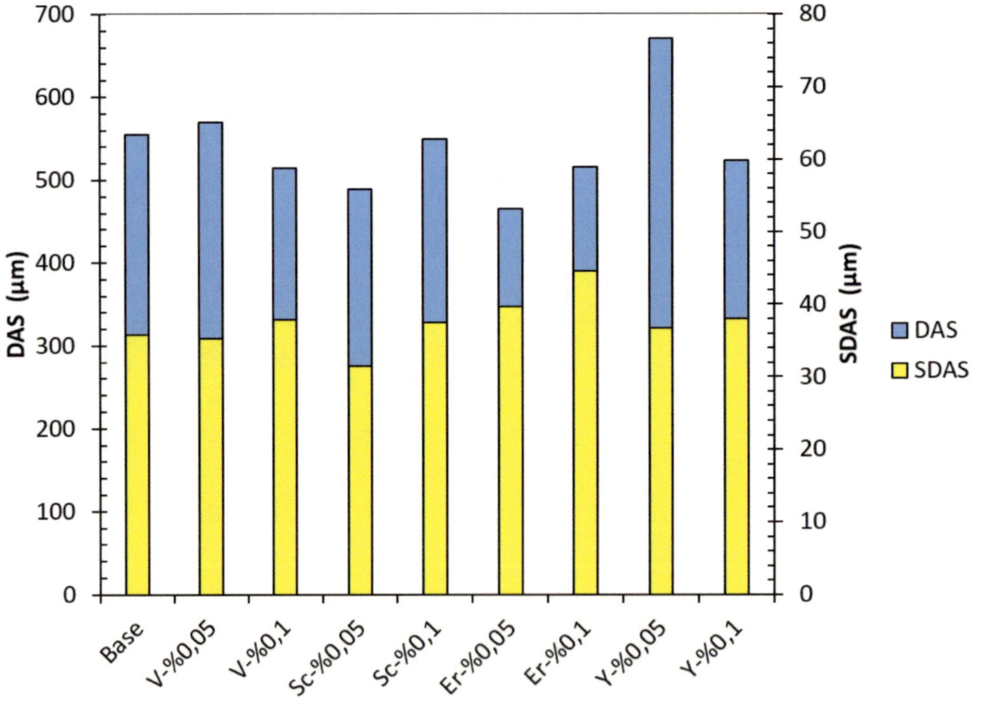

Figure 12.
DAS and SDAS change of different alloying element additions.

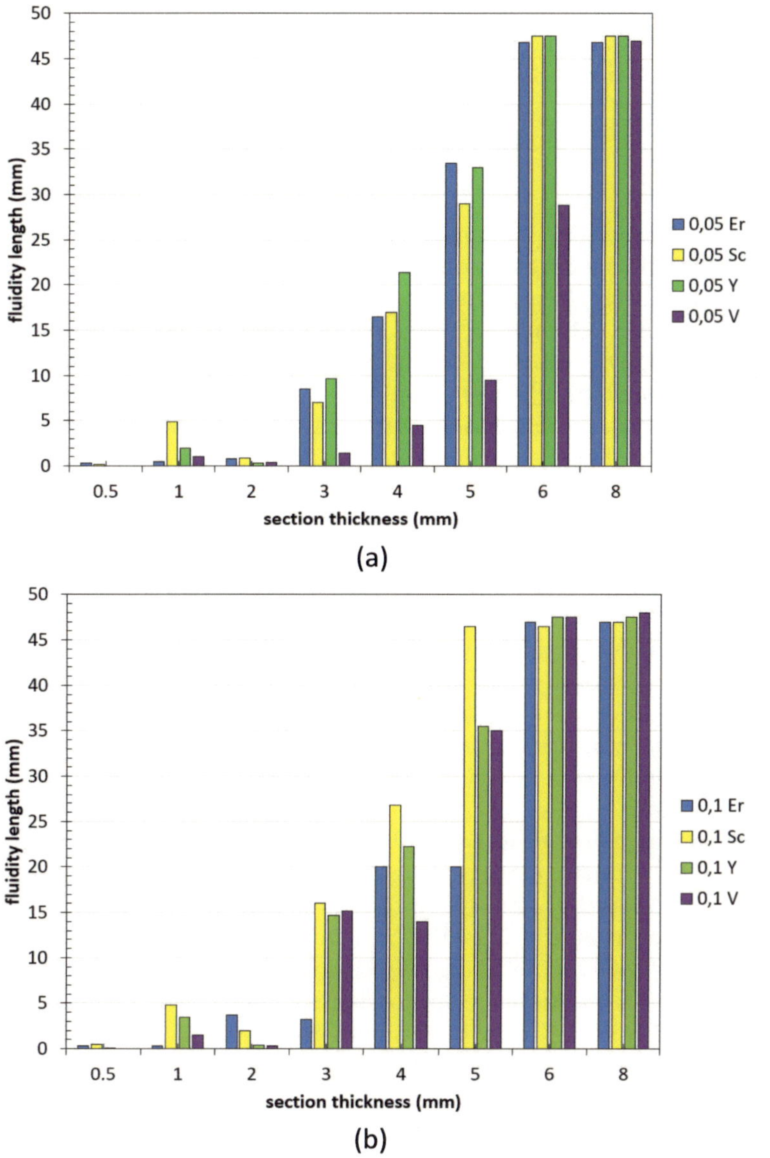

Figure 13.
Fluidity change of different grain refiner (a) 0.05 wt%, (b) 0.1 wt% additions.

together with a 30% reduction in SDAS. However, in this work, 0.05 and 0.1 wt% Er were added and no modification of Si was observed (**Figures 10c** and **11c**).

Pramod [22] reported that 0.4 wt% Sc had a reduction capability of SDAS by 50%. Compare to 0.2 wt% Sc addition, the silicon size was also significantly decreased which resulted in having the highest UTS values of 300 MPa. Pandee [38] added three different levels of Sc to A356; namely 0.24, 0.40, and 0.65 wt%. Two different cooling rates were studied. At both cooling rate levels (0.1 and 3°C/s), as Sc content was increased, Si morphology was modified to a fibrous structure. At 3°C/s, the modification of Si was found to be the optimum for 0.4 wt% Sc modified A356. In a similar kind of work, W. Zhang [4] also reported similar findings that higher than 0.15 wt %Sc addition to A356, Si morphology was completely modified.

On the other hand, Xu [25] found almost a linear decrease in grain size with increased Sc content varying from 0.2 towards 0.8 wt% Sc additions to F357 where the modification of Si was started at 0.2 wt% additions. With UTS values as high as 350 MPa.

3.3 Fluidity

The fluidity test results were analyzed under the wt% additions as seen in **Figure 13**. When 0.05 wt% addition levels are compared, it can be seen that Er and Y appear to have higher fluidity lengths compared to the other additions. On the other hand, when 0.1 wt% additions are analyzed, Sc has significantly higher fluidity compared to the other additions. In each of the cases, V reveals the lowest fluidity results. Prukkanon [39] reported that up to 0.2 wt% Sc addition, the fluidity of A356 was increased, but a higher amount of Sc addition had not affected the fluidity and remained unchanged at 660, 690, and 720°C.

4. Conclusion

The addition of 0.1 wt% Sc exhibits the highest fluidity in all cross-sections ranging from 0.5 to 8 mm thicknesses whereas V addition shows the lowest fluidity.

Although, the microstructural changes are not so significantly different from each other (all additions have SDAS values in the range of 30–40 μm), in terms of tensile properties, Sc addition to A357 alloy reveals the highest reliable and reproducible results. On the other hand, Y additions have the highest scatter with being the lowest reliable.

Before heat treatment, almost all additions have yield stress in the range of 119–130 MPa and ultimate tensile strength between 180 and 210 MPa. The addition of 0.1 wt% Sc has the highest UTS of 218 MPa which is quite closely followed by 0.1 wt% Er with a value of 212 MPa. Similarly, elongation at fracture values lie between 3.6 and 4.6% where Y is the lowest and Sc and Er is the highest.

After heat treatment, there are approximately two folds of increase in the yield strength for all additions. Y is the lowest with 210 MPa, and Sc and Er are the highest with 260 and 240 MPa, respectively. There is an approximately a 30% increase in UTS after heat treatment. Er is the highest with 300 MPa, followed by Sc at 299 Mpa. Y is the lowest with 240 Mpa. Elongation at fracture values were decreased dramatically after heat treatment from 4% towards 1%. Er shows the highest elongation with 4.9% while Y is the lowest with 1.2%.

In almost all additions, toughness value decreased after heat treatment, however, only when 0.1 wt% Er was added to A357, there was no difference in the toughness before and after heat treatment.

Author details

Kerem Can Dizdar[1], Hayati Sahin[1], Furkan Tezer[1] and Derya Dispinar[1,2*]

1 Metallurgical and Materials Engineering, Istanbul Technical University, Istanbul, Turkey

2 NonFerrous Metal Treatment, R&D Center, Foseco, Enschede, Netherlands

*Address all correspondence to: derya.dispinar@gmail.com

IntechOpen

© 2022 The Author(s). Licensee IntechOpen. This chapter is distributed under the terms of the Creative Commons Attribution License (http://creativecommons.org/licenses/by/3.0), which permits unrestricted use, distribution, and reproduction in any medium, provided the original work is properly cited.

References

[1] Samuel A, Elgallad E, Mahmoud M, et al. Rare earth metal-based intermetallics formation in Al–Cu–Mg and Al–Si–Cu–Mg alloys: A metallographic study. Advances in Materials Science and Engineering. 2018;**2018**:1-15

[2] Jia K, Yu W-B, Yao J-M, et al. Al–9.00% Si–0.25% Mg alloys modified by ytterbium. Rare Metals. 2017;**36**:95-100

[3] Yun D, Zheng R, Xiaoping L, et al. Investigation on the modification behavior of A356 alloy inoculated with a Sr-Y composite modifier. Journal of Rare Earths. 2013;**31**:204-208

[4] Zhang W, Liu Y, Yang J, et al. Effects of Sc content on the microstructure of As-Cast Al-7 wt.% Si alloys. Materials Characterization. 2012;**66**:104-110

[5] Li B, Wang H, Jie J, Wei Z. Effects of yttrium and heat treatment on the microstructure and tensile properties of Al–7.5 Si–0.5 Mg alloy. Materials and Design. 2011;**32**:1617-1622

[6] Li J, Wang X, Ludwig T, et al. Modification of eutectic Si in Al–Si alloys with Eu addition. Acta Materialia. 2015;**84**:153-163

[7] Mazahery A, Habibnejad-korayem M, Takrouri K. Performance of europium in microstructural modification of high strength lightweight aluminum components. International Journal of Lightweight Materials and Manufacture. 2019;**2**:250-254

[8] Mao F, Yan G, Xuan Z, et al. Effect of Eu addition on the microstructures and mechanical properties of A356 aluminum alloys. Journal of Alloys and Compounds. 2015;**650**:896-906

[9] Li LF, Li DQ, Luo M, et al. Influence of Rare Earth Additions on the Microstructure and Mechanical Properties of Al7Si0.3Mg Alloys Processed by Semi-Solid Die Casting and Gravity Die Casting. Trans Tech Publ. 2019. pp 69-74

[10] Xiong J, Yan H, Zhong S, Bi M. Effects of Yb addition on the microstructure and mechanical properties of as-cast ADC12 alloy. Metals. 2019;**9**:108

[11] Kabliman E, Bozorgi S, Kronsteiner J, Nikolas A. Development of Al–Mg–Sc thin foils for fiber-reinforced metal laminates. Advanced Engineering Materials. 2019;**21**:1800462

[12] Elgallad E, Ibrahim M, Doty H, Samuel F. Microstructural characterisation of Al–Si cast alloys containing rare earth additions. Philosophical Magazine. 2018;**98**: 1337-1359

[13] Zhang S, Leng J, Wang Z, et al. Investigation on the modification behavior of A356. 2 alloy with Yb–La composite modifier. Materials Research Express. 2018;5:016520

[14] Colombo M, Gariboldi E, Morri A. Er addition to Al-Si-Mg-based casting alloy: Effects on microstructure, room and high temperature mechanical properties. Journal of Alloys and Compounds. 2017;**708**:1234-1244

[15] Shi Z, Wang Q, Zhao G, Zhang R. Effects of erbium modification on the microstructure and mechanical properties of A356 aluminum alloys. Materials Science and Engineering A. 2015;**626**:102-107

[16] Gao Z, Li H, Lai Y, et al. Effects of minor Zr and Er on microstructure and mechanical properties of pure aluminum. Materials Science and Engineering A. 2013;**580**:92-98

[17] Pengfei X, Bo G, Zhuang Y, et al. Effect of erbium on properties and microstructure of Al-Si eutectic alloy. Journal of Rare Earths. 2010;**28**:927-930

[18] Mahmoud M, Elgallad E, Ibrahim M, Samuel F. Effect of rare earth metals on porosity formation in A356 alloy. International Journal of Metalcasting. 2018;**12**:251-265

[19] Zhang X, Wang Z, Zhou Z, Xu J. Influence of rare earth (Ce and La) addition on the performance of Al-3.0 wt% Mg alloy. Journal Wuhan University of Technology-Materials Science Edition. 2017;**32**:611-618

[20] Ahmad R, Asmael M, Shahizan N, Gandouz S. Reduction in secondary dendrite arm spacing in cast eutectic Al–Si piston alloys by cerium addition. International Journal of Minerals, Metallurgy, and Materials. 2017;**24**:91-101

[21] Pourbahari B, Emamy M. Effects of La intermetallics on the structure and tensile properties of thin section gravity die-cast A357 Al alloy. Materials and Design. 2016;**94**:111-120

[22] Pramod S, Rao AP, Murty B, Bakshi SR. Effect of Sc addition and T6 aging treatment on the microstructure modification and mechanical properties of A356 alloy. Materials Science and Engineering A. 2016; **674**:438-450

[23] Yii SJ, Anas N, Ramdziah M, Anasyida A. Microstructural and mechanical properties of Al-20% Si containing cerium. Procedia Chemistry. 2016;**19**:304-310

[24] Tzeng Y-C, Jian S-Y. Effects of the addition of trace amounts of Sc on the microstructure and mechanical properties of Al-11.6Si alloys. Materials Science and Engineering A. 2018;**723**:22-28. DOI: 10.1016/j.msea.2018.03.016

[25] Xu C, Xiao W, Hanada S, et al. The effect of scandium addition on microstructure and mechanical properties of Al–Si–Mg alloy: A multi-refinement modifier. Materials Characterization. 2015;**110**:160-169

[26] Campbell J, Harding R. TALAT Lecture Notes: The Fluidity of Molten Metals. Lecture 3205. 1994

[27] Colak M, Kayikci R, Dispinar D. Influence of different cross sections on fluidity characteristics of A356. Transactions of the Indian Institute of Metals. 2015;**68**:275-281

[28] Sjölander E, Seifeddine S. The heat treatment of Al–Si–Cu–Mg casting alloys. Journal of Materials Processing Technology. 2010;**210**:1249-1259

[29] Hu X, et al. Effects of rare earth Er additions on microstructure development and mechanical properties of die-cast ADC12 aluminum alloy. Journal of Alloys and Compounds. 2012;**538**:21-27

[30] Dispinar D, Campbell J. Critical assessment of reduced pressure test. Part 1: Porosity phenomena. International Journal of Cast Metals Research. 2004;**17**:280-286

[31] Dispinar D, Campbell J. Porosity, hydrogen and bifilm content in Al alloy castings. Materials Science and Engineering A. 2011;**528**:3860-3865

[32] Dispinar D, Campbell J. Critical assessment of reduced pressure test. Part 2: Quantification. International Journal of Cast Metals Research. 2004;**17**:287-294

[33] Dispinar D, Campbell J. Use of bifilm index as an assessment of liquid metal quality. International Journal of Cast Metals Research. 2006;**19**:5-17

[34] Dispinar D, Campbell J. Effect of casting conditions on aluminium metal

quality. Journal of Materials Processing Technology. 2007;**182**:405-410

[35] Dispinar D, Akhtar S, Nordmark A, et al. Degassing, hydrogen and porosity phenomena in A356. Materials Science and Engineering A. 2010;**527**:3719-3725

[36] Uludağ M, Çetin R, Dispinar D, Tiryakioğlu M. Characterization of the effect of melt treatments on melt quality in Al-7wt% Si-Mg alloys. Metals. 2017;**7**:157

[37] Tiryakioğlu M, Campbell J. Quality index for aluminum alloy castings. International Journal of Metalcasting. 2014;**8**:39-42

[38] Pandee P, Gourlay CM, Belyakov SA, et al. $AlSi_2Sc_2$ intermetallic formation in Al-7Si-0.3Mg-xSc alloys and their effects on as-cast properties. Journal of Alloys and Compounds. 2018;**731**:1159-1170. DOI: 10.1016/j.jallcom.2017.10.125

[39] Prukkanon W, Srisukhumbowornchai N, Limmaneevichitr C. Influence of Sc modification on the fluidity of an A356 aluminum alloy. Journal of Alloys and Compounds. 2009;**487**:453-457

Chapter 5

Selection of Optimal Material from Stir Cast Aluminum Graphene Nano Platelets Composites for Aerospace Applications

Bhanu Prakash Palampalle, Babu Dharmalingam and Devika Royal

Abstract

Qualitative and quantitative requirements when selecting materials for different properties can be difficult and ambiguous. An insufficient variety of materials can lead to component malfunction and failure at any point during their service. Owing to the vast availability of dissimilar materials, material selection in the engineering design phase is difficult and elusive. This study presents an EDAS (Evaluation based on Distance from Average Solution) and VIKOR (VIse Kriterijumska Optimizacija-kompromisno Resenje) techniques for effective material selection for aviation applications. In this research, the selection index value was calculated using the EDAS and VIKOR entropy-based weight techniques. The MADM (multi-attribute decision making) procedure also selects the best weight per cent combination among pure aluminum reinforced with GNPs (graphene nanoplatelets) for aircraft applications based on its physical and mechanical properties. The results demonstrate that 0.5 wt% GNPs reinforced in pure aluminum has the best combination of both physical and mechanical qualities, according to the EDAS and VIKOR multi-criteria decision-making methodologies. The composites were made using the stir casting technique. MATLAB R2020a is used to grade and compare the composite materials.

Keywords: pure aluminum, multi-attribute decision making, graphene nanoplatelets, VIKOR, EDAS, MATLAB R2020a

1. Introduction

Aluminum Metal Matrix Composites are favored over other conventional materials in aerospace, automotive, and marine applications because of improved properties such as high strength-to-weight ratio, good wear resistance, and so on. Graphene's mechanical, physical, optical, and thermal properties make it an outstanding metal composite reinforcement material. Pure aluminum graphene nanoplatelets (GNPs) were reinforced in a base matrix (pure Al) with different weight percentages to form aluminum metal matrix composites using stir casting [1, 2] powder metallurgy [3] and other techniques. The uniform distribution of graphene nanoplatelets [4, 5] in the aluminum matrix improves mechanical properties significantly. Stephen et al. [6] discovered that graphene-aluminum nanocomposites had lower strength and stiffness than pure aluminum reinforced with multi-walled carbon nanotube composites due to the production of enhanced aluminum carbide with the graphene filler. Venkata Subbaiah et al. [7] investigated the microstructural and mechanical properties of AA7075-GNPs. In an aluminum 7075 base matrix, ultrasonic-assisted stir casting was used to produce graphene nanoplatelets varying from 0.5 to 2.0 wt%. The composite with 0.5% GNPs had the highest tensile strength and microhardness due to the less porosity and uniform distribution of GNPs in the AA7075 matrix. According to Bhanu Prakash et al. [8], the microstructure and mechanical properties of aluminum 7075 graphene nanoplatelets ranged from 0.50 to 2 wt% in base matrix fabricated using the stir casting technique. In comparison with other weight percentages, AA 7075–1.5% GNPs provided the composite with better mechanical properties. Due to the graphene's uniform distribution with the base metal, Muhammed Emre Turan et al. [9] discovered that adding graphene to pure magnesium in different weight proportions improved hardness values. Xin Gao et al. [10] stress the contents of graphene reinforced with base minerals. The effect of graphene on the tensile strength of the prepared composites increases by 0.3 wt% as the graphene content increases. As the proportion of graphene in the composite increases, the tensile strength and percentage of elongation to fracturing decrease. A systematic approach to GNP composites' selection is necessary to choose the best material for a given application. The correct material selection technique entails precisely describing the application requirement in terms of mechanical properties, primarily for the utility class defined in the proposed application. Various researchers have used MADM methods such as VIKOR, EDAS, WSM PROMETHE, and TOPSIS to select the best material for specific applications in a range of fields such as automotive [11–16], marine [17, 18], medical [19], and agriculture [20–22]. The VIKOR method outperformed the other 10 most common methods for selecting suitable materials for a sailing boat mast, a flywheel, and a cryogenic storage tank, according to the author [23]. The optimal material, according to the researcher [24], is solely determined by the criterion's maximum priority value. The most conclusive of the three MCDM methods is VIKOR (TOPSIS, VIKOR, and PROMETHEE). Caliskan et al. [25] rated the materials using the PROMETHEE II, TOPSIS, and VIKOR methods and compared the results obtained by each process. Tungsten carbide-cobalt and Fe-5Cr-Mo-V aircraft steel were found to be the best materials for tool holder production. A new version of the VIKOR method, based on criteria for selecting the best material, particularly in the biomedical field, was proposed by Jahan et al. [26]

2. Problem description and experimental details

In this research, the best material for aircraft application was chosen from five options of aluminum graphene nanoplatelet composites. **Figure 1** depicts the beneficial and non-beneficial criteria. MADM's EDAS and VIKOR methods are used to choose the right option. Procedural steps for criterion methods are represented in **Figure 2**.

2.1 Composite fabrication

Stir casting has been used for the manufacturing of pure aluminum GNPs composites because it has a lower initial cost than other fabrication techniques. The author detailed the manufacturing of Al-GNPs composites in his prior work [1]. Mechanical stirring was used to distribute the reinforcing phases in the molten matrix metal during the fabrication process. **Table 1** displays the matrix and reinforcing materials used in the composite fabrication process.

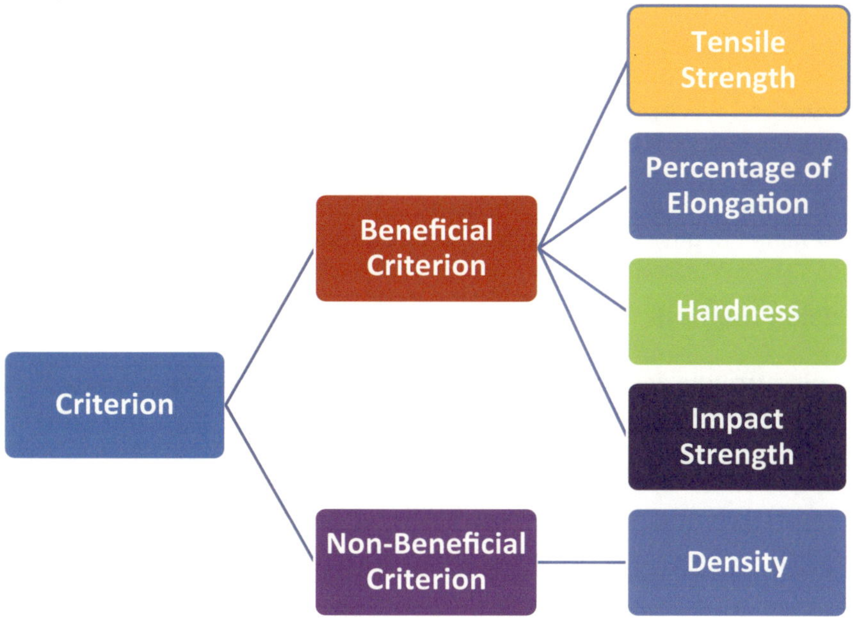

Figure 1.
Beneficial and non-beneficial criterion.

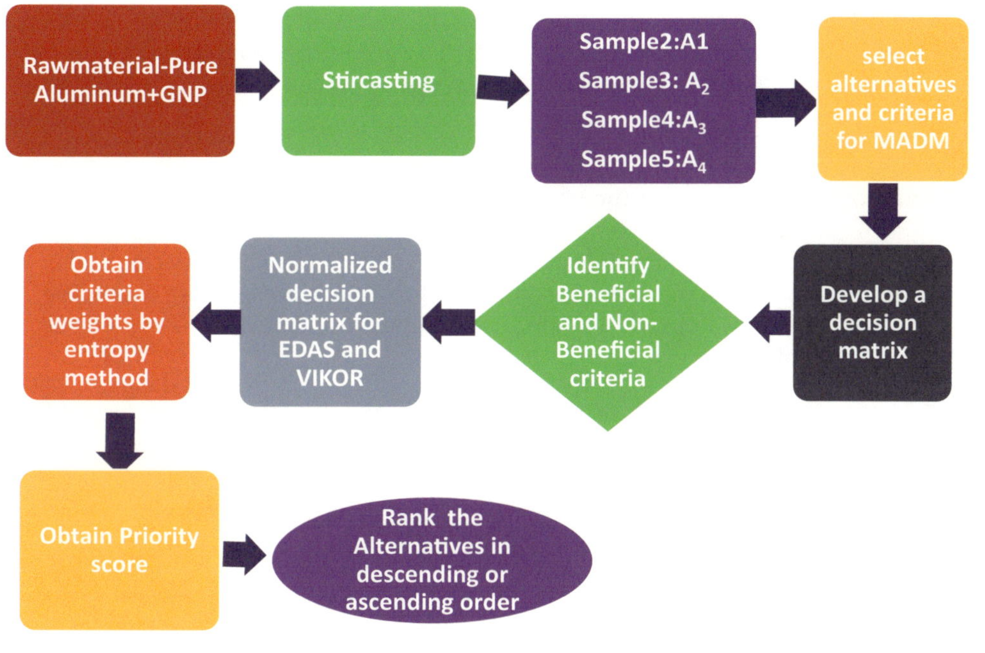

Figure 2.
Applied steps for EDAS and VIKOR.

Sl. no.	Matrix materials	Reinforcements	Size of reinforcements
1	Pure aluminum (99.5%)	GNPs (0.5–2.0 wt.%)	2–10 nm

Table 1.
Material and reinforcement for composite fabrication.

2.2 Characterization of a composite of pure aluminum and graphene nanoplatelets

Physical and mechanical tests on the fabricated composite were used to classify the pure aluminum GNPs composite. These experiments were carried out on well-prepared specimens prepared according to ASTM Standards such as ASTM E8 & ASTM A370 for Tensile Test and Impact test, to investigate the impact of reinforcements in pure aluminum for aircraft applications.

3. Decision methods

A multi-attribute decision-making problem is represented by a matrix X, which contains n alternatives and m criteria.

$$X = \begin{pmatrix} X_{11} & X_{12} & X_{13} & - & - & - & X_{1m} \\ X_{21} & X_{22} & X_{23} & - & - & - & X_{2m} \\ X_{31} & X_{32} & X_{33} & - & - & - & X_{3m} \\ X_{41} & X_{42} & X_{43} & - & - & - & X_{4m} \\ - & - & - & - & - & - & - \\ - & - & - & - & - & - & - \\ X_{n1} & X_{n2} & X_{n3} & - & - & - & X_{nm} \end{pmatrix} \tag{1}$$

where X_{ij} is the ith alternative's output on the jth criterion. In MADM methods, the weight value (w_j) for each criterion must be determined such that the number of all criterion weights equals to one. The entropy approach is used to calculate these weights.

3.1 Evaluation based on distance from average solution (EDAS) method

The EDAS method is a distance-based technique that employs positive and negative distances from the average solution. The options are listed in ascending order. The procedural steps for n alternative composite and m parameters suggested [27, 28] are as follows:

$$\text{Let } [X]_{n*m} \text{ be the decision} - \text{making matrix} \tag{2}$$

Step 1. The average solution (AV_j).

$$AV_j = \sum_{i=1}^{n} X_{ij}/n \tag{3}$$

Step 2. The positive distance from average (PDA) for beneficial and non-beneficial criterion.

$$\left.\begin{array}{c} \text{If } j^{th} \text{ criterion is beneficial,} \\ PDA_{ij} = \dfrac{\max\left(0, \left(X_{ij} - AV_j\right)\right)}{AV_j} \\ \text{And if } j^{th} \text{criterion is non-beneficial,} \\ PDA_{ij} = \dfrac{\max\left(0, \left(AV_j - X_{ij}\right)\right)}{AV_j} \end{array}\right\} \tag{4}$$

Step 3. The negative distance from average (NDA) for beneficial and non-beneficial criterion.

$$\left.\begin{array}{c}\text{If } j^{th} \text{ criterion is beneficial,}\\ NDA_{ij} = \dfrac{\max\left(0, (AV_j - X_{ij})\right)}{AV_j}\\ \text{And if } j^{th} \text{criterion is non-beneficial,}\\ NDA_{ij} = \dfrac{\max\left(0, (X_{ij} - AV_j)\right)}{AV_j}\end{array}\right\} \tag{5}$$

Step 4. Eq. (6) calculates the positive and the negative weighted sums.

$$SP_i = \sum_{j=1}^{m} \left(W_j * PDA_{ij}\right) \tag{6}$$

$$SN_i = \sum_{j=1}^{m} \left(W_j * NDA_{ij}\right) \tag{7}$$

W_j = weight of jth criterion.

Step 5. Normalized values of SP_i and SN_i.

$$NSP_i = \frac{SPi}{max\, i\, SPi} \tag{8}$$

$$NSN_i = \frac{SNi}{max\, i\, SNi} \tag{9}$$

Step 6. The appraisal score (AS_i) for all alternatives is calculated using Eqs. (8) and (9).

$$AS_i = 0.5 * (NSPi + NSN_i) \tag{10}$$

Where $0 \le AS_i \le 1$.

The alternative with the outstanding appraisal score is chosen as the best, among the other selective alternatives.

3.2 VIse Kriterijumska Optimizacija kompromisno Resenje (VIKOR) method

According to the VIKOR technique given by Rao [29] and Chatterjee et al. [30] for material selection, the compromise may be common for resolving the dispute, and the practicable option may be nearest to the ideal solution. The options are primarily assessed based on all of the characteristics taken into account.

Step 1. Determine the decisive criteria and select the best alternatives based on those criteria.

Step 1: (a) In the decision matrix, determine the best, (x_{ij}) max and the worst, (x_{ij}) min values of all the criteria.

$$X_i^+ = \max{}_i\left(X_{ij}\right)[J = 1, 2 \ldots, m] \tag{11}$$

$$X_i^- = \min{}_i\left(X_{ij}\right)[J = 1, 2 \ldots, m] \tag{12}$$

Step 2: Compute the values of utility measure (S_i) and regret measure (R_i).

$$S_i = \sum_{j=1}^{m} \left(W_j^* \left((X_i^+ - X_{ij}) / (X_i^+ - X_i^-)\right)\right) \tag{13}$$

$$R_i = \max_j\left(W_j^*\left(\left(X_i^+ - X_{ij}\right)/\left(X_i^+ - X_i^-\right)\right)\right) \quad (14)$$

Step 3: Find the values of S^*, S^-, R^*, R^-.

$$S* = \min_i(S_i) \; S^- = \max_i(S_i) \quad (15)$$

$$R* = \min_i(R_i) \; R^- = \max_i(R_i) \quad (16)$$

Step 4. Compute the value of Q_i for j = 1, 2, ... , m.

$$Q_i = N*((S_i - S*)/(S^- - S*)) + (1 - N)*((R_i - R*)/(R^- - R*)) \quad (17)$$

The alternatives are ranked in the ascending order for the Q_i values. The one having the lower value of Q_i is considered as the best alternative.

4. Results and discussions

Table 2 displays the decision matrix for the pure aluminum graphene nanoplatelets (GNP's) composites, which contains five alternatives and five parameters (1 & 2). The entropy method was used to calculate the objective weights, which are described in **Table 3**. MATLAB R2020a software was used to create a programme that used the formulae for the methods (EDAS and VIKOR) to determine performance measures. The entropy methods are used to measure the weights of different parameters such as mass, percentage elongation, tensile strength (TS), hardness, and impact strength (IS). The material that is used for fuselage construction is of lightweight and good strength and is more corrosion resistant. An attempt was made by developing an alternative material to serve the purpose.

4.1 Evaluation based on distance from average solution (EDAS) method

Table 1 decision matrix displays the average solution (AVj) determined using Eq. (3). Eqs. (4) and (5) were used to quantify positive and negative distances based

S. no	Al + wt.% GNPs	Density (g/cc)	% Elongation	UTS (MPa)	Micro-hardness	Impact strength (kJ/m^2)
1	Pure Al	2.68	38	77	25	127
2	Pure Al + 0.5	2.68	64	88	56	850
3	Pure Al + 1.0	2.68	56	85	44	775
4	Pure Al + 1.5	2.68	38	77	38	750
5	Pure Al + 2.0	2.68	35	65	35	735
AV_J		2.681	46	79	39	647

Table 2.
Decision matrix.

Weight	Density	%Elongation	UTS	Micro-hardness	Impact strength
W	0.000005	0.170289	0.028746	0.183363	0.617597

Based on the Entropy method, weights were calculated and used for criterion selection.

Table 3.
Entropy weights of criteria.

S. no	Al + GNP wt%	Density	% Elongation	UTS	Hardness	IS
1	Pure Al	0	0	0	0	0
2	Pure Al + 0.5	0.00067	0.38977	0.12573	0.41989	0.31311
3	Pure Al + 1.0	0.00067	0.21625	0.08064	0.1034	0.19725
4	Pure Al + 1.5	0.00067	0	0	0	0.15863
5	Pure Al + 2.0	0.00067	0	0	0	0.13546

Table 4.
(PDA) positive distance from average.

S. no	Al + Wt.%GNP	Density	% Elongation	UTS	Micro-hardness	Impact strength
1	Pure Al	0.00268	0.18176	0.01806	0.36702	0.80445
2	Pure Al + 0.5	0	0	0	0	0
3	Pure Al + 1.0	0	0	0	0	0
4	Pure Al + 1.5	0	0.18176	0.01806	0.03484	0
5	Pure Al + 2.0	0	0.24251	0.17025	0.12143	0

Table 5.
(NDA) negative distance from average.

S. no	Al + wt% GNP	SP_i	SN_i	NSP_i	NSN_i	AS_i	Rank
1	Pure Al	0.000000	0.595597	0	0	0	5
2	Pure Al + 0.5	0.340359	0.000000	1	1	1	1
3	Pure Al + 1.0	0.179926	0.000000	0.52864	1	0.76432	2
4	Pure Al + 1.5	0.097970	0.037858	0.28784	0.93644	0.61214	3
5	Pure Al + 2.0	0.083658	0.068456	0.24579	0.88506	0.56543	4

Table 6.
Normalized positive and negative weighted sums, appraisal scores, and rank of alternatives.

on the types of beneficial and non-beneficial parameters mentioned in **Tables 4** and 5. Eqs. (6) and (7) are used to measure the positive and negative weighted sums (7). Eqs. (8) and (9) are used to measure the normalized positive and negative weighted sums (9). Eq. (10) is used to measure the final assessment score (AS_i) for all alternatives, as shown in **Table 6**. 5-1-2-3-4 is the ranking of alternative composite materials. The best composite for the fuselage in aircraft applications is pure aluminum reinforced with 0.5wt% GNP, though pure aluminum is the least favored.

4.2 VIse Kriterijumska OptimizacijaKompromisno Resenje (VIKOR) method

The related utility and regret measures, as well as Qi values for the five alternative composite materials and the parameters, are determined using Eqs. (13)–(17), as shown in **Table 7** (taking N = 0.5). **Table 8** shows the best and worst values for the beneficial and non-beneficial criteria measured using Eqs. (11) and (12). The ranks of alternative composite materials are 5-1-2-3-4. The first place winner is pure

S. no	Al + wt% GNP	S_i	R_i	Q_i	Rank
1	Pure Al	0.96886	0.6176	1	5
2	Pure Al + 0.5	0	0	0	1
3	Pure Al + 1.0	0.18889	0.07375	0.15718	2
4	Pure Al + 1.5	0.35922	0.15393	0.31	3
5	Pure Al + 2.0	0.42335	0.17029	0.35634	4

Table 7.
Utility and regret measure values and rank of the alternatives.

Type	Density	% Elongation	UTS	Micro-hardness	Impact strength
Best X_i^+	2.68	64.075	88.39	56.08	850
Worst X_i^-	2.689	34.924	65.15	25	126.58

Table 8.
Best and worst of criterion.

aluminum reinforced with 0.5 wt% GNPs. VIKOR is more beneficial compared to EDAS, though the results are comparatively same. All the expressions were objective with respective to VIKOR to EDAS.

5. Conclusion

The EDAS and VIKOR methods were used in this research to assist in the selection of the best alternative composite material for aircraft applications. The study was performed on pure aluminum graphene nanoplatelets composites manufactured using the stir casting technique. As shown in **Figure 3**, decision-making using EDAS and VIKOR methodology using MATLAB R2020a shows that out of the five alternatives, alternative 2 (pure Al + 0.5 wt% GNP) is the best suitable material for fuselage construction in the aerospace industry.

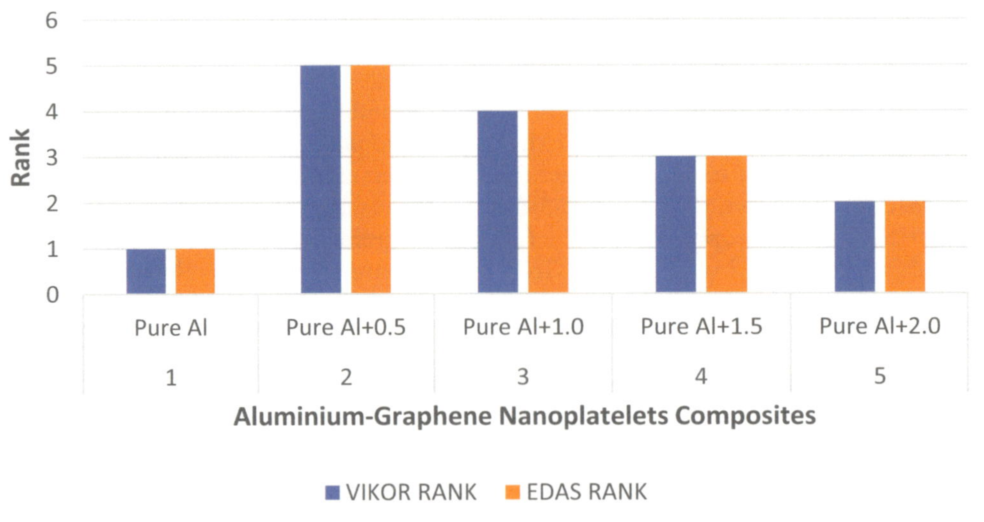

Figure 3.
Comparison of ranks.

Nomenclature

Symbols

X_i^+, X_i^-	ideal best and worst solutions
SP_I, SN_i	positive and negative weighted sums
NSP_I, NSN_i	normalized weighted sums
AS_i	appraisal score

Acronyms

MADM	multi-attribute decision making
AHP	analytical hierarchy process
TOPSIS	Technique for Order of Preference by Similarity to Ideal Solution
VIKOR	VIse Kriterijumska OptimizacijaKompromisno Resenje
EDAS	Evaluation based on Distance from Average Solution
PROMETHEE	Preference Ranking Organization Method for Enrichment Evaluation
ELECTRE	ELimination and Choice Expressing REality
UTS	ultimate tensile strength
IS	impact strength
GNPs	graphene nanoplatelets

Author details

Bhanu Prakash Palampalle[1]*, Babu Dharmalingam[1] and Devika Royal[2]

1 Centre for Research and Innovation, K.S.R.M. College of Engineering, Kadapa, India

2 Department of Mathematics, Gitam University, Bengaluru, India

*Address all correspondence to: bhanupalem@ksrmce.ac.in; bhanupalem@gmail.com

IntechOpen

© 2021 The Author(s). Licensee IntechOpen. This chapter is distributed under the terms of the Creative Commons Attribution License (http://creativecommons.org/licenses/by/3.0), which permits unrestricted use, distribution, and reproduction in any medium, provided the original work is properly cited.

References

[1] Subbaiah V et al. Microstructural analysis and mechanical properties of pure Al–GNPs composites by stir casting method. Journal of the Institution of Engineers (India): Series C. 2019;**100**:493-500. DOI: 10.1007/s40032-018-0491-1

[2] Venkatesan S et al. Mechanical behaviour of aluminium metal matrix composite reinforced with graphene particulate by stir casting method. Journal of Chemical and Pharmaceutical Sciences. 2017;**1**:55-59

[3] Rashad M et al. Investigation on micro structural, mechanical and electro chemical properties of aluminium composites reinforced with graphene nano platelets. Progress in Natural Science: Materials International. 2015; **25**:460-470

[4] Yolshina LA et al. Enhancement of the mechanical properties of aluminium–graphene composites. American Institute of Physics. 2016; **1785**:040093

[5] Shin SE et al. Deformation behaviour of aluminium alloy matrix composites reinforced with few-layer graphene. Composites Part A: Applied Science and Manufacturing. 2015;**78**:42-47

[6] Bartolucci SF et al. Graphene–aluminium nano composites. Materials Science and Engineering A. 2011;**528**: 7933-7937

[7] Subbaiah KV et al. Microstructural analysis and mechanical behaviour of aluminium-GNPs composites fabricated by ultrasonic assisted stir casting technique. IJMPERD. 2018;**8**(8):385-392. ISSN (P): 2249-6890; ISSN (E): 2249-8001

[8] Prakash PB et al. Microstructure analysis and evaluation of mechanical properties of Al 7075 GNP's composites. Materials Today: Proceedings. 2018;**5**: 14281-14291

[9] Turan ME et al. The effect of GNPs on wear and corrosion behaviours of pure magnesium. Journal of Alloys and Compounds. 2017;**724**:14-23

[10] Gao X et al. Preparation and tensile properties of homogeneously dispersed graphene reinforced aluminium matrix composites. Materials and Design. 2016; **94**:54-60

[11] Dev S, Aherwar A, Patnaik A. Material selection for automotive piston component using entropy-VIKOR method. Silicon. 2019;**12**(1):155-169

[12] Moradian M, Modanloo V, Aghaiee S. Comparative analysis of multi criteria decision making techniques for material selection of brake booster valve body. Journal of Traffic and Transportation Engineering. 2019;**6**(5):526-534

[13] Sen B, Bhattacharjee P, Mandal U. A comparative study of some prominent multi criteria decision making methods for connecting rod material selection. Perspectives on Science. 2016;**8**:547-549

[14] Giorgetti A, Cavallini C, Arcidiacono G, Citti P. A mixed C-VIKOR fuzzy approach for material selection during design phase: A case study in valve seats for high performance engine. International Journal of Applied Engineering Research. 2017;**12**(12):3117-3129

[15] Sasanka CT, Ravindra K. Implementation of VIKOR method for selection of magnesium alloy to suit automotive applications. International Journal of Advanced Science and Technology. 2015;**83**(5):49-58

[16] Girubha RJ, Vinodh S. Application of fuzzy VIKOR and environmental impact analysis for material selection of an automotive component. Materials and Design. 2012;**37**:478-486

[17] Gangwar S, Arya P, Pathak VK. Optimal material selection for ship body based on fabricated zirconium dioxide/silicon carbide filled aluminium hybrid metal alloy composites using novel fuzzy based preference selection index. Silicon. 2021;**13**:2545-2562. DOI: 10.1007/s12633-020-00600-4

[18] Yadav S, Pathak VK, Gangwar S. A novel hybrid TOPSIS-PSI approach for material selection in marine application. Sådhanå. 2019;**44**(58):1-12

[19] Bahraminasab M and Jahan A Material selection for femoral component of total knee replacement using comprehensive VIKOR Materials and Design. September 2011;**32**(8–9): 4471-4477. DOI: 10/1016/j.matdes.2011.03.046

[20] Sudha AS. Application of EDAS method on water requirement in agriculture. International Journal of Engineering Research and Technology. 2019;**8**(12):558-561

[21] Sowmya Dhanalakshmi C, Madhu P, Karthick A, Mathew M, Kumar RV. A comprehensive MCDM-based approach using TOPSIS and EDAS as an auxiliary tool for pyrolysis material selection and its application. Biomass Conversion and Biorefinery. 2020. DOI: 10.1007/s13399-020-01009-0, ISSN: 2190-6815

[22] Mitta A. Selection of cotton fabrics using EDAS method. Journal of Natural Fibers. 2020:1-13. DOI: 10.1080/15440478.2020.1821289

[23] Athawale VM, Chakraborty S. Material selection using multi-criteria decision-making methods: A comparative study. Proceedings of the Institution of Mechanical Engineers, Part L: Journal of Materials: Design and Applications. 2012;**226**(4):266-285. DOI: 10.1177/1464420712448979

[24] Chakraborty S, Chatterjee P. Selection of materials using multi-criteri decision making methods with minimum data. Decision Science Letters. 2013;**2**(3):135-148

[25] Çalıskan H, Kursuncu B, Kurbanoğlu C, Güven SY. Material selection for the tool holder working under hard milling conditions using different multi criteria decision making methods. Materials and Design. 2013;**45**: 473-479

[26] Jahan A, Mustapha F, Ismail M, Sapuan S, Bahraminasab M. A comprehensive VIKOR method for material selection. Materials and Design. 2011;**32**:1215-1221

[27] Chatterjee P, Banerjee A, Mondal S, Boral S, Chakraborty S. Development of a hybrid meta-model for material selection using design of experiments and EDAS method. Engineering Transactions. 2018;**66**(2):187-207

[28] Ghorabaee MK, Amiri M, Zavadskas EK, Turskis Z, Antucheviciene J. Stochastic EDAS method for multi-criteria decision-making with normally distributed data. Journal of Intelligent and Fuzzy Systems. 2017;**33**(3):1627-1638. DOI: 10.3233/JIFS-17184

[29] Rao RV. A decision making methodology for material selection using an improved compromise ranking method. Materials and Design. 2008; **29**(10):1949-1954

[30] Chatterjee P, Athawale VM, Chakraborty S. Selection of materials using compromise ranking and outranking methods. Materials and Design. 2009;**30**(10):4043-4053. DOI: 10.1007/s40032-018-0491-1

Chapter 6

Assisting Liquid Phase Sintering of Pure Aluminum (Al) by the Tin Addition

Nur Ayuni Jamal, Farazila Yusof, Yusilawati Ahmad, Norhuda Hidayah Nordin and Suraya Sulaiman

Abstract

In the present study, the addition of tin (Sn) to the pure Al system was done, and its effects on the morphology, density, and compressive yield strength of pure Al were analyzed systematically. In this context, the morphology of sintered Al revealed enhanced wettability and sintering response between Al particles with increased Sn content. Moreover, physical characteristics of sintered Al alloys demonstrated oxidation phenomenon (black color specimen) with the lowest Sn content of 1.5 weight percent (wt.%), in which a higher Sn content of 2 and 2.5 wt.% produced silver color specimens, implying a reduction in oxidation. Additionally, densification of sintered Al alloys was greatly promoted with increased Sn contents, suggesting effective wetting as confirmed by the previous morphological observations. Similarly, the compressive yield strength of sintered Al alloys improved with increased Sn content which might be due to the enhanced inter-particle contacts between Al particles and sufficient wetting by molten Sn. Based on the results obtained, the introduction of Sn powder at various contents improved the sintering response of pure Al powder by providing sufficient liquid-phase sintering. Therefore, the sintered Al alloys had enhanced the morphological, densification, physical characteristics, and compressive yield strength.

Keywords: aluminum alloys, tin, liquid phase sintering, powder metallurgy, oxidation, sintered density, compressive strength

1. Introduction

Lightweight materials, especially aluminum (Al) and its alloys, are increasingly utilized in different industries because of their uniqueness, including high corrosion resistance, outstanding strength to weight ratio, excellent flexibility, thermal properties, and remarkable finishing properties [1–3]. As such, Al and its alloys are finding increased use in the fields of automotive, construction, marine, and aerospace. Solid-state and liquid-state processes are widely implemented to fabricate metals to obtain Al and its alloys that could be applied in the abovementioned fields [1, 4–7]. However, there is an increased interest in the solid-state process and practice due to its versatility, greater control of starting materials proportion, chemical homogeneity, potential to fabricate near net shape with complex parts, and cost-effectivity [5–8]. The studies by Ji et al. [8] and Feijoo et al. [9] documented a noteworthy improvement in the ultimate tensile strength (UTS) of around 180 megapascals (MPa) and 377 MPa of an Al alloy system with an optimum reinforcement addition of 5 weight percent (wt.%) silicon carbide (SiC) and 0.5 wt.% multiwall carbon nanotubes, fabricated through powder metallurgy technique.

However, despite its advantages, the sintering of Al and its alloys is problematic due to a thermodynamically stable oxide layer on the particle surfaces. The scenario often hinders effective bond formations between the powder particles. Therefore, the accomplishment of desirable physical and mechanical performances could not be realized. Moreover, the presence of the unwanted oxide layer frequently promotes a nonwetting diffusion barrier for effective sintering of Al and its alloys [10–16]. Consequently, it is crucial to disrupt at least partially the oxide film, known as aluminum oxide (Al_2O_3) or alumina, on the surfaces of Al particles. Therefore, the addition of sintering additives such as tin (Sn), zinc (Zn), and magnesium (Mg) to facilitate the liquid-phase sintering of Al and its alloys were introduced to minimize the issues [11–19]. Azadbeh and Razzaghi [15] reported that when 5.5 wt.% of Zn was introduced to an Al system, its densification and mechanical strength were improved at 119 HV and 564 MPa, respectively. Furthermore, a study performed by Liu et al. [16] documented that the addition of Sn (4 wt.%) could only provide sufficient liquid formation for effective liquid phase sintering of Al after the oxide layer on the Al surface has been interrupted via external load application or Mg addition. In this context, the authors recorded that the Al-Sn liquid development proceeded via melting of Sn followed by a repetition process of cracking and repairing owing to reoxidation and thermally induced stress of the exposed Al particles. Consequently, as the outer layer of Al particles scavenged the oxygen from the Argon (Ar) gas, the oxygen concentration in the core was reduced, accompanied by the creation of an Al-Sn liquid phase via the dissolution of the Al in the liquid Sn. Finally, sacrification of outer layers took place hence these layers remained porous. Additionally, the authors also suggested that an alternative mechanism by means of selection of irregular shape of Sn element could ensure maximum liquid formation during sintering of Al.

Therefore, the sintering of the Al cores proceeded through a typical liquid-phase sintering mechanism. Similarly, the utilization of sintering additives (Mg and Sn) enhanced the sintering response of sintered Al, which consequently improved its tensile strength up to 118 and 300 MPa, as discussed in the studies by Sercombe et al. [13] and MacAskill et al. [17]. Although there were extensive works on the sintering response of Al and its alloys fabricated via powder metallurgy technique, additional exploration is necessary, particularly with different formulations of Al and sintering additive. Consequently, the behavior of liquid-phase sintering of Al alloys was investigated in the current study by exploiting Sn constituent as sintering additive, developed through powder metallurgy technique. Moreover, Mg constituent at fixed content was also utilized in the current study to optimize the liquid phase formation by Sn constituent. Therefore, the reliability of the sintering response of Al alloys was systematically examined via its physical characteristics, oxygen content levels, compressive strength, density, and microstructure of the sintered Al alloys. The investigation aimed to provide an alternative solution to minimize the problematic sintering response of Al and its alloys.

2. Experimental procedure

The base material, Al powder with 99.9% purity was purchased from NovaScientific (M) Sdn Bhd. The sintering additives to facilitate liquid phase sintering of Al utilized elemental powders of Sn (99.5% purity) and Mg (99.9% purity) were supplied by NovaScientific (M) Sdn Bhd and Sigma Aldrich (M) Sdn Bhd, respectively. The information on the purity of these starting materials was provided by the respective suppliers. To observe the behavior of liquid phase sintering of Al alloys, varied content of Sn between 1.5, 2, and 2.5 wt.% and a constant 1 wt.% of Mg was added. Moreover, pure Al without the addition of Sn and Mg constituents was set as the reference material prior to study the effects of Sn and Mg in assisting liquid phase sintering of Al. As schematically illustrated in **Figure 1**, a powder metallurgy technique of mixing, compaction, and sintering was implemented to fabricate a sintered Al alloy. In this process, table-top ball milling was employed to prepare the elemental powder mixture comprised of Al, Sn, and Mg. The mixing was conducted for 2 h with the ball to powder ratio of 1 to 10, where zirconia balls were used as the ball milling medium to produce a homogenous elemental powder mixture. Consequently, the elemental powder mixture was pressed into cylindrical pellets of 10 mm diameter and 6 mm height under a constantly applied pressure of 250 MPa. Then, the compacted specimens were sintered under a protected environment of argon (Ar) gas at 580°C for 2 h to obtain pure sintered Al alloys. The sintered alloys were then cleaned with acetone to remove impurities, followed by drying in an oven for 12 h to prepare them for characterization. The microstructure of as received starting materials and elemental powder mixture were analyzed under scanning electron microscopy (SEM) (SEM, Jeol JSM6500F, JEOL Ltd., Tokyo, Japan) before they were employed to fabricate the sintered Al alloy. The step was essential to develop satisfactory sintered Al alloys

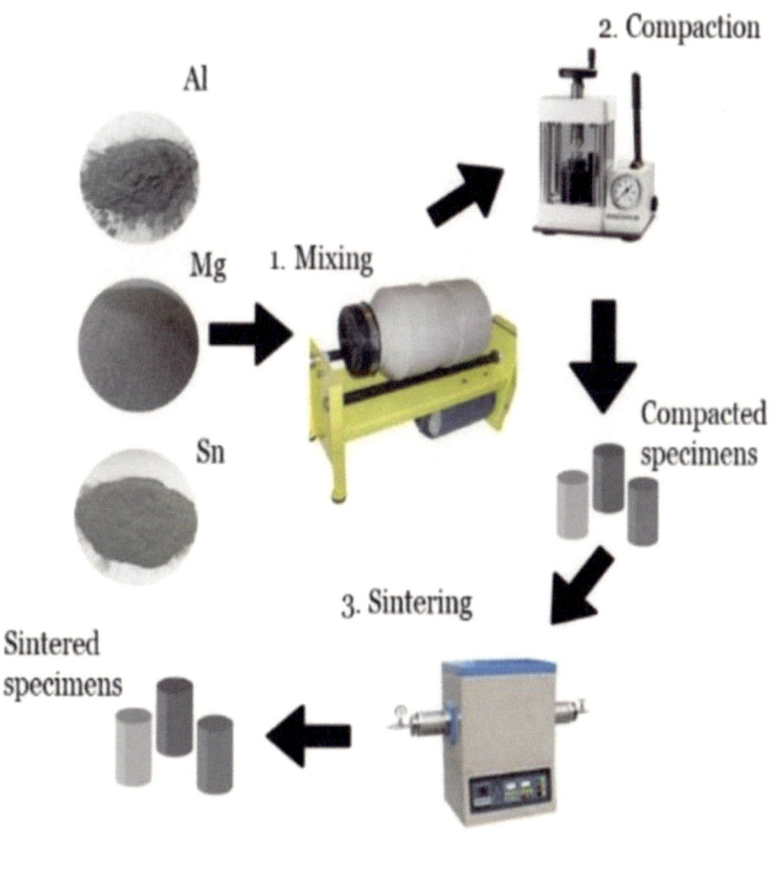

Figure 1.
Schematic diagram of the development of Al alloys by powder metallurgy technique.

with desirable performance. A cross-section of sintered Al alloys was prepared for microstructure observation through SEM. Collectively, the residual oxygen content of the Al alloys at every stage was monitored by energy dispersive spectroscopy (EDS). Furthermore, any possibility of phase changes of the elemental powder mixture and the sintered Al alloys were identified using x-ray diffraction (XRD) analysis (XRD; PANanalytical empyrean 1032, Eindhoven, Netherlands) in the 2θ range of 20–80°. Moreover, the step size of 0.05 degrees, radiation source of Cu Kα and accelerating voltage and current of 40 kV and 40 mA were also set for the XRD test conditions. Finally, the sintered density of the resultant Al alloys was determined via the Archimedes principle, as per ASTM C830-93. Similar experimental procedure was performed in fabricating the reference material (pure Al).

3. Results and discussion

3.1 Microstructure of the starting materials and elemental powder mixture

The microstructure of the starting materials and elemental powder mixture is illustrated in **Figure 2(a–c)** and **Figure 3(a–c)**. From the observations, most of the Al and Mg particles were spherical, with some irregularities observed on the Al particles. These particles were calculated to have an average particle size of 45 and 10 μm, respectively. On the contrary, the shape of the Sn particles was found to be mainly irregular with an average particle size of 45 μm, as revealed in **Figure 2(c)**. Since the sintering of Al powder prevented the solid-state sintering process, it was crucial to disrupt the stable

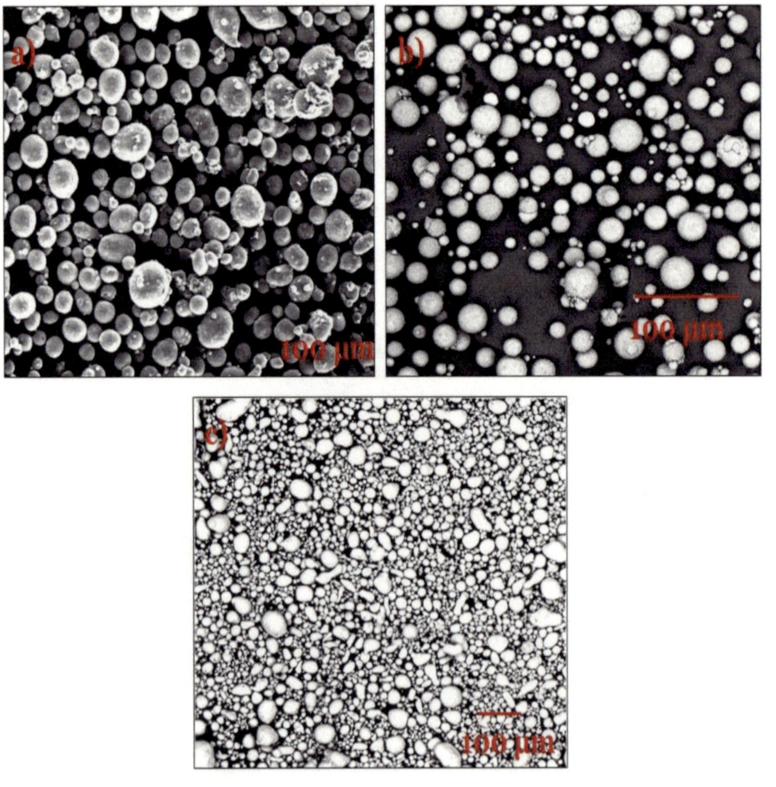

Figure 2.
Microstructure of the starting materials of (a) Al powder, (b) Mg powder, and (c) Sn powder as observed under SEM.

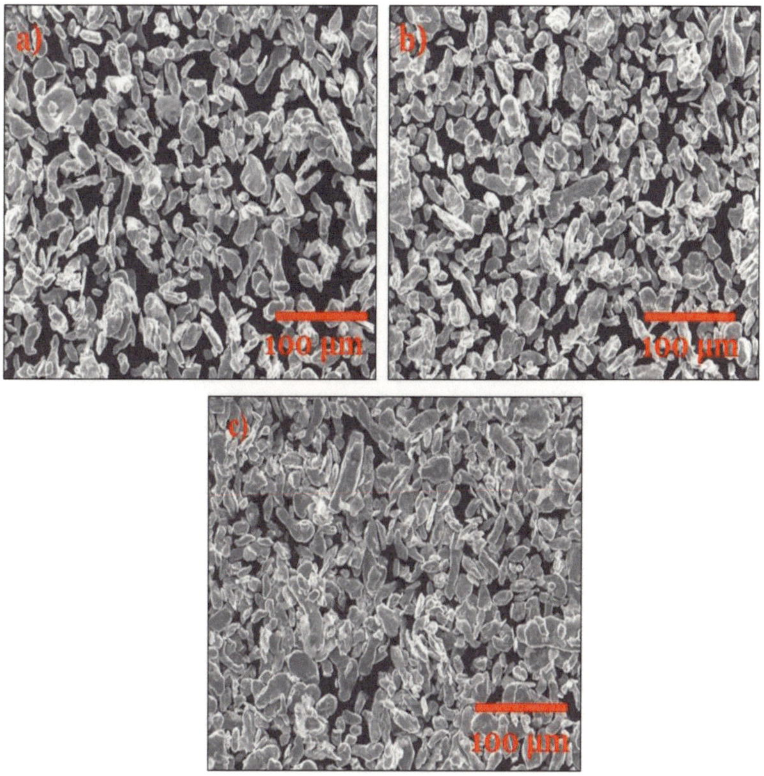

Figure 3.
Microstructure of elemental powder mixture at different Sn content of (a) 1.5 wt.%, (b) 2 wt.%, and (c) 2.5 wt.% as observed under SEM.

oxide film that readily existed on the surface of the Al particles so that strong metallurgical bonds between the particles could be guaranteed. Therefore, the current investigation employed liquid phase sintering of Al alloys with Sn as the sintering additive. Sn was introduced to promote the liquid phase of Al because of its low melting temperature (232°C) compared to Al (660°C). However, the process could only be successful in the presence of Mg. The addition of Mg was reported to reduce the surface oxide of Al by exposing the underlying metal thus resulting in possible wetting of the Al particles by liquid Sn [14, 18–19]. On the other hand, the absence of powder mixture accumulation and homogenous distribution were identified after 12 h of mixing, suggesting optimal mixing condition in preparing elemental powder mixture of Al, Sn, and Mg, as demonstrated in **Figure 2(d)**. Regardless of different Sn content, the resultant particles of the elemental powder mixture were found to exhibit a lamellar structure, as evidenced in **Figure 3(a–c)**, respectively. These Sn particles were also identified to be flattened and elongated, having an average particle size of 50 µm. The resultant microstructure of the elemental powder mixture could be due to the repeated welding, fracturing, and rewelding of the elemental powder particles during mixing via ball milling technique [20]. Nevertheless, the overlapping and intimate contact between Sn particles were also clearly visible with increasing Sn content, confirming an increased Sn content in the elemental powder mixture, as seen in **Figure 3(a–c)**, respectively.

3.2 Visual inspection of the sintered aluminum alloys

The visual inspection of the sintered Al alloys at various Sn contents are depicted in **Figure 4(a–c)**. It was observed that the addition of 1.5 wt.% of Sn

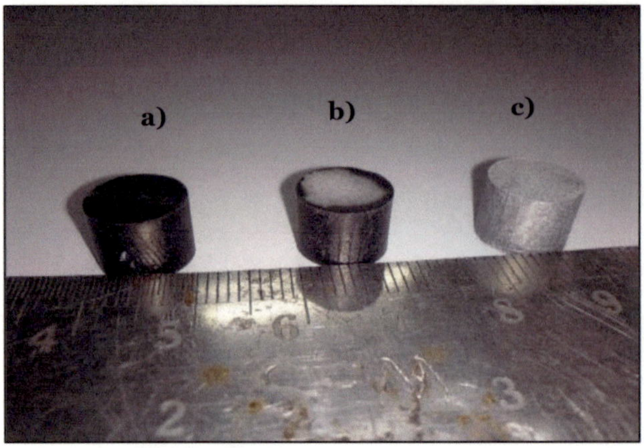

Figure 4.
Visual inspection of the resultant Al alloys with different Sn content of (a) 1.5 wt.%, (b) 2 wt.%, and (c) 2.5 wt.%.

(the lowest amount) produced a black-colored Al alloy after sintering, implying oxidation and the ineffective role of Sn in increasing the fluidity of the Al alloy during sintering. However, when the Sn content was increased to 2 wt.%, the black-colored Al alloy changed slightly to silver that appeared only at the top of the alloy, suggesting insufficient liquid formation to enhance the Al alloy fluidity that consequently minimizing oxidation. Similar observation was also reported in the study of Pan et al. [21]. On contrary, the effects were more pronounced when the Sn content was maximized to 2.5 wt.% as the color of Al alloys completely replicated the original silver color of the Al powder, confirming complete interruption of the stable oxide film on the surfaces of Al during sintering. Furthermore, the findings were supported by the results obtained from the oxygen content analysis as tabulated in **Table 1**. A significantly decreased oxygen content of the resultant Al alloys from 0.58 to 0.44 wt.% was recorded when the content of Sn was increased from 1.5 to 2.5 wt.%, respectively. The results showed that the amount of Sn influenced the oxygen content of the resultant Al alloys. Therefore, as the Sn content was increased in the Al alloys, the oxygen level decreased, indicating an improvement of the overall wetting characteristics of the Al alloys. However, the addition of Mg constituent was also essential in assisting the formation of Sn liquid by reducing the oxide layer on the Al surface [4, 18–19]. According to Kondoh et al. [14], Mg addition was found to assist the distribution of Sn liquid below the oxide layer, thus contributing to successful wetting of Al by Sn liquid.

3.3 Microstructure of the resultant aluminum alloys

The cross-section of the sintered Al alloys at 1.5, 2, and 2.5 wt.% of Sn are demonstrated in **Figure 5(a–c)**, respectively. Moreover, the cross section of the pure Al without Sn and Mg addition (reference sample) is shown in **Figure 5(d)**. Based on the observations, the reference sample exhibited poor sintering response; evidenced by visible appearance of individual particles as if no sintering process has been performed. Furthermore, the presence of large number of pores between the individual particles were obvious, confirming ineffective sintering of pure Al. Similarly, pores or inter-particle voids between the grain boundaries were also noticeable when the Sn amount in the Al system was the lowest (1.5 wt.%), portraying a poor sintering response. However, the microstructure of 1.5 wt.% containing Al was slightly improved in comparison to pure Al without Sn addition.

Tin content (wt.%)	Oxygen content (wt.%)
Elemental powder mixture	0.22 ± 0.10
1.5 (sintered at 580°C)	0.58 ± 0.12
2 (sintered at 580°C)	0.51 ± 0.13
2.5 (sintered at 580°C)	0.44 ± 0.11

Table 1.
Oxygen content reading of elemental powder mixture and sintered Al alloys at various Sn content from EDS analysis. Data are presented in mean ± standard deviation.

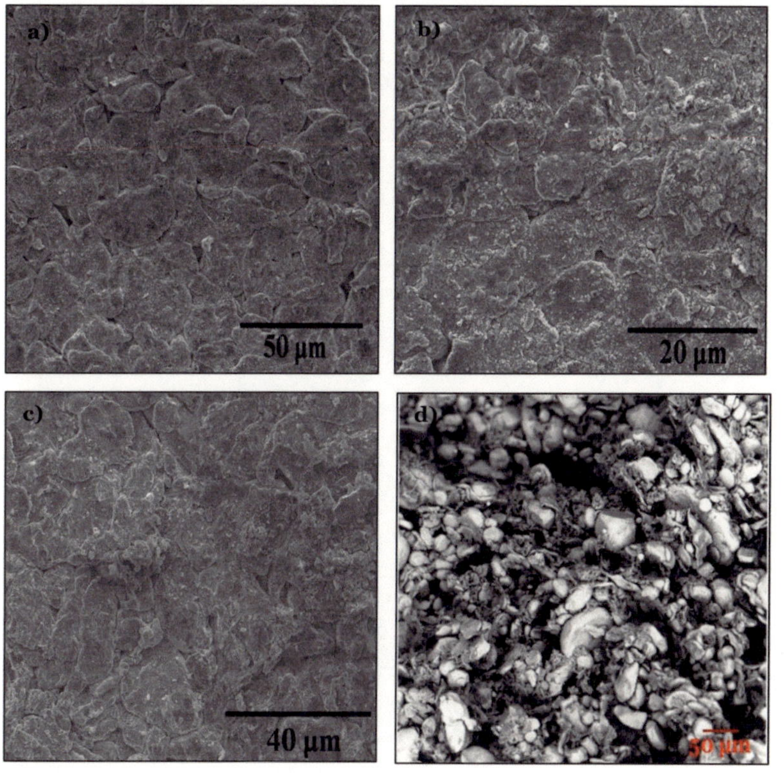

Figure 5.
Microstructures of the resultant Al alloys at various Sn content of (a) 1.5 wt.%, (b) 2 wt.%, (c) 2.5 wt.%, and (d) pure Al without Sn and Mg constituents (reference sample).

Therefore, the results might be associated with insufficient pore rounding as evidenced by the appearance of elongated pores, an abundance of grain boundaries, and negligible inter-particle necking [17–19, 22]. Moreover, the addition of Sn might be inadequate to occupy both the elongated and rounded pores, preventing a more favorable packing grains arrangement that could bring the solid particles closer, which contributed to grain growth and densification failure. On the contrary, improved sintering response was observed with increased Sn content, particularly at 2 and 2.5 wt.%, owing to the increased presence of pores rounding and lesser particle boundaries. Additionally, grain growth development was observed with increased Sn content as the amount of liquid formation was sufficient to fill the existing pores that successfully brought the solid particles together, improving the sintering quality of the resultant Al specimens. According to Aneta [23], a desirable liquid phase enhanced mass transport, which encouraged the rearrangement and fragmentation of solid particles. The phenomenon was especially true as the formation of inter-particle necking became increasingly visible with increased pore rounding, as shown in **Figure 5(b** and **c)**.

Furthermore, successful wetting of Sn elements on the surfaces of the Al particles enhanced the metallurgical bonds between the Al particles, attributable to higher and sufficient liquid phase formation that filled the remaining pores in the Al system [17–19, 22, 24]. Therefore, the addition of higher Sn content produced a permanent liquid phase that improved the densification and sintering quality of the alloys (mainly for Al added with 2 and 2.5 wt.% of Sn). Undetectable clustering of Sn particles represented by the bright phases indicated a uniform distribution of Sn particles along the grain boundaries of the Al alloys, especially with the addition of 2 and 2.5 wt.% of Sn that further promoted effective liquid phase formation. Consequently, densification and compressive strength of the Al alloys were elevated, as evidenced in **Figures 6** and 7, respectively. According to MacAskill et al. [17], such morphologies confirmed the liquid formation by molten Sn during sintering that was capable of wetting the surface of Al and scattered evenly

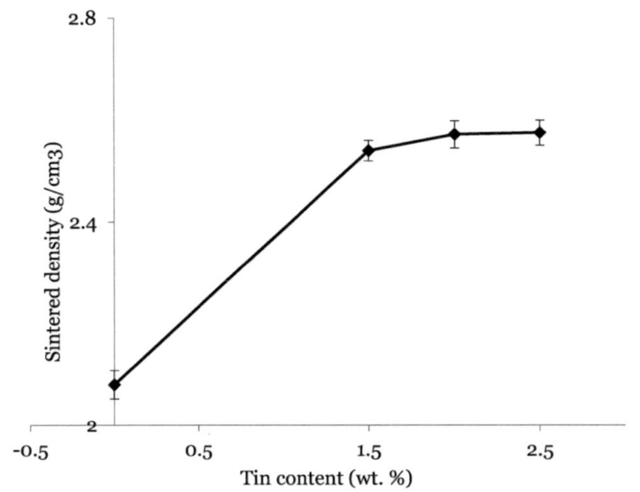

Figure 6.
The effect of Sn variation on the sintered densities of the resultant Al alloys.

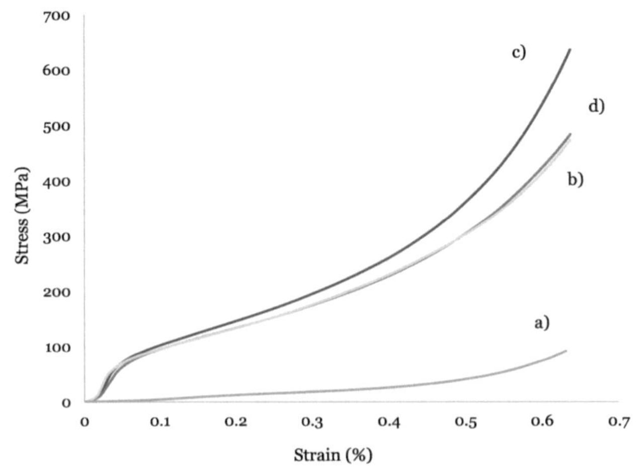

Figure 7.
Stress-strain curve of resultant (a) pure Al and Al alloys with different Sn content of (b) 1.5 wt.%, (c) 2 wt.%, and (d) 2.5 wt.%.

throughout the Al alloys. However, the occurrence could only be accomplished by the existence of Mg due to its potential to stimulate the wetting response of molten Sn. However, higher Sn content of up to 2.5 wt.% slightly reintroduced several particle boundaries as shown in **Figure 4(c)**. The observation might be due to the remaining unreacted Sn that was unable to wet the Al surfaces, which hindered effective liquid phase formation during sintering [25]. Moreover, Sercombe et al. [13] reported a similar observation.

3.4 Phases transformation of the resultant aluminum alloys

The X-ray diffraction (XRD) patterns of pure Al without Sn and Mg addition as well Al alloys at different Sn content (1.5, 2, and 2.5 wt.%) are presented in **Figure 8(a–d)**. The dominant XRD phases of the resultant Al alloys were mainly attributed to the α-Al constituent, characterized by the (111), (200), (220), and (311) diffraction peaks at 38.87, 45.42, 67.16, and 78.54°, respectively. As can be seen in **Figure 8(a)**, the peak intensities of pure Al were found to be lower compared to the 1.5, 2, and 2.5 wt.% Sn containing Al alloys. This is probably due to poor sintering response in the absence of Sn and Mg constituents hence low crystalline formation during complete sintering. In the case of Al alloys, the peak intensities of Sn increased with increased Sn content, confirming the addition of Sn during the process. The absence of the Mg peak in the Al system was attributed to the dissolution of Mg in α-Al [17, 26]. Based on the Al-Mg binary equilibrium phase diagram, Mg addition up to maximum content of 18.6% could be completely dissolved in the parent Al phase at the eutectic temperature of 450°C [27]. Moreover, it has been

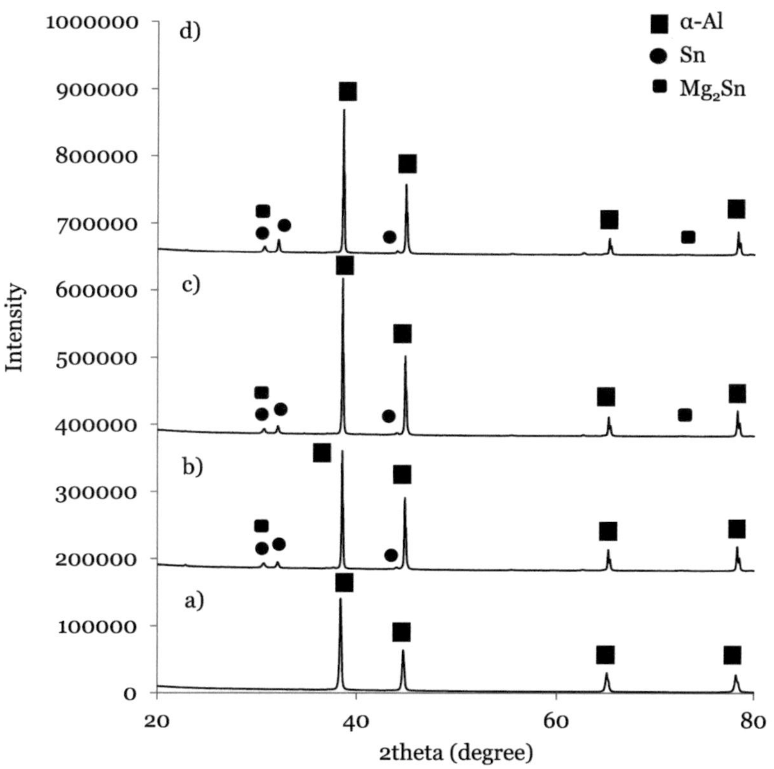

Figure 8.
Phases transformation of (a) Pure Al and Al alloys with different Sn content of (b) 1.5 wt.%, (c) 2 wt.%, and (d) 2.5 wt.%.

reported that the creation of brittle intermetallic compounds (β-Al_3Mg_2 phase) that has associated with poor mechanical performance was observable with increasing Mg content [28]. For this reason, its utilization should be maximized up to 5 and 10 wt.% for wrought alloys and cast alloys, respectively [28]. Moreover, it was documented that Mg vaporized during sintering, and the gas produced served as a reliable gettering agent for effective liquid phase sintering of Al [14, 26, 29–30]. Therefore, it has been proposed that the application of nitrogen (Ni) as a sintering atmosphere was effective in transporting Mg vapor (effective gettering agent) around the pore network via the formation of magnesium nitride (Mg_3N_2) [29]. On the contrary, Schaffer and Hall [29] documented that Mg could be effectively transported as a vapor in Ar atmosphere which is comparable to the Ni atmosphere considering higher vapor pressure of Mg as compared to Mg_3Ni_2 formation. On the other hand, weak peaks of Mg_2Sn compounds were observed in the current study with a higher Sn amount (2 and 2.5 wt.%), postulating solidification of the Mg_2Sn compounds within the Al phase [31]. This is essential for optimum attainment of physical and mechanical performance of the resultant alloys [31]. It is important to note that a lower sintering temperature of 580°C that was set in the current study prevented the formation of higher peaks of the Mg_2Sn compounds. Therefore, future studies should analyze the effects of various sintering temperatures other than the current setting on the sintering response of Al alloys.

3.5 Sintered densities of the resultant aluminum alloys

Figure 6 presents the sintered densities of pure Al and 1.5, 2, and 2.5 wt.% Sn containing Al alloys. Based on the findings, the densification of pure Al was found to enhance from 2.083 g/cm^3 to 2.5397 g/cm^3, 2.573 g/cm^3 and 2.575 g/cm^3, with the addition of Sn content from 1.5 to 2 wt. % and 2.5 wt.%, respectively. A comparable observation was documented in previous literature [23–24, 26, 30]. The low sintered density value of pure Al might be due to the noticeable appearance of individual particles and large number of pores formation hence portraying poor sintering response as evidenced in **Figure 5(d)**. In contrast, the addition of a higher amount of Sn reduced micro pores formation and diminished particle boundaries, which improved the sintered densities as shown in **Figure 5(b–d)**. Moreover, the lessened appearance of elongated pores might have promoted the densification effects in the current study. According to Ozay et al. [24], enhanced densification might be the result of filled gaps between the powder particles through the action of molten Sn, which also contributed to reduced porosity. Furthermore, Sercombe et al. [13] suggested that improved wetting effects with the implementation of higher Sn content led to reduced surface tension as Sn dispersed into the liquid phase during sintering. Atabay et al. [31] also highlighted that the existence of Mg_2Sn intermetallic compounds contributed to the positive effects of densification of the Al alloys. Therefore, the highest sintered density of the Al alloys was observed when 2.5 wt.% of Sn was added. This is further supported by the higher and more intense visible XRD peaks of the intermetallic compounds compared to the pure Al and Al alloys with 1.5 wt.% of Sn as demonstrated in **Figure 7(c)**.

3.6 Compressive strength of resultant aluminum alloys

A compressive test was conducted to evaluate the response of the pure Al and 1.5, 2, and 2.5 wt.% Sn containing Al alloys under a compressive load. The stress-strain curve of the resultant pure Al and Al alloys with varying Sn content is shown in **Figure 7**. The obtained stress-strain curve of the pure Al and Al alloys

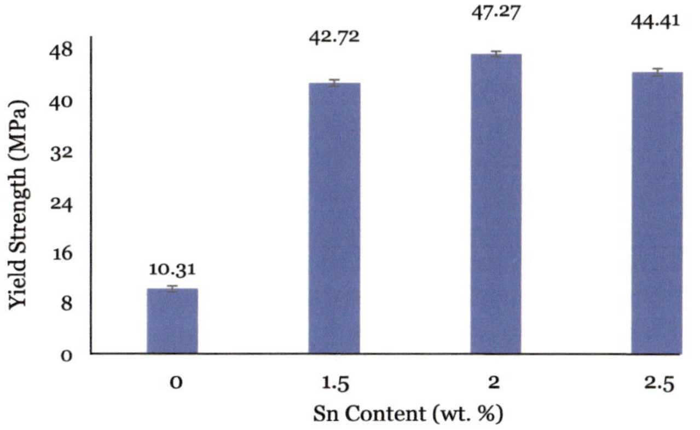

Figure 9.
Compressive yield strength of pure Al and Al alloys with different Sn content of 1.5, 2, and 2.5 wt.%.

obeyed the typical stress-strain curve of Al and its alloys. To better understand, the compressive yield strength of the pure Al and Al alloys is displayed in **Figure 9**. Based on the findings, the compressive yield strength of the pure Al demonstrated the lowest yield strength value of 10.31 MPa, implying poor sintering response owing to the presence of individual particles with greater number of pores formation as seen in **Figure 5(d)**. This shows that the absence of Sn as sintering additive in combination with Mg constituent was unfavorable in promoting the liquid phase sintering of pure Al. On the other hand, Al alloys exhibited an increasing trend from 42.72 to 47.27 MPa with the Sn content of 1.5 and 2 wt.%, respectively. However, the addition of 2.5 wt.% Sn (the highest) reduced the compressive yield strength of the Al alloys from 47.27 (2 wt.% Sn) to 44.21 MPa. Therefore, the results were inconsistent with the results from the sintered density measurements, presumably due to a slightly increased particle boundary or grain coarsening, as revealed in **Figure 5(c)**. Furthermore, parallel observations were also discussed by Padmavathi and Upadhyaya [26].

Although the presence of Mg_2Sn compounds (**Figure 8(b** and **c)**) produced substantial improvements in the mechanical performance of Al and its alloys, the effects of grain coarsening were more pronounced, which reduced the compressive yield strength of the Al alloys when the Sn content was maximized to 2.5 wt.% [31]. Moreover, it was reported that excessive content of Sn caused undiffused liquid phase to remain at the particle boundary and led to the brittle creation of particle boundary networks, that consequently reduced the mechanical performance of Al and its alloys [13, 31, 17–19]. Therefore, it could be hypothesized that particle boundary enlargement and liquid phase profusion might contribute to the decreased compressive strength of the Al alloys that were fabricated in the current study.

4. Conclusion

The utilization of Sn element that served as sintering additive in the current study significantly improved the sintering response of the resultant Al alloys that were fabricated via powder metallurgy technique. The varying Sn content between 1.5, 2, and 2.5 wt.% enhanced the properties of the resultant Al alloys in terms of their physical appearance, from black to sliver-like coloring due to oxidation-reduction, and sintered densities from 2.5397 to 2.575 g/cm^3 as the gaps between particles were filled by molten Sn, which lessened pore formation and appearance of particle boundary. Therefore, the microstructure of the resultant Al alloys confirmed the pore rounding and diminished particle boundary that led to the improved metallurgical bonding between the Al particles.

Additionally, the compressive yield strength of the resultant Al alloys increased from 42.72 to 47.27 MPa with an increased amount of Sn (from 1.5 to 2 wt.%), which were the result of diminished particle boundary and reduced pores formation. However, despite a positive increment in the sintered density of the Al alloys with increased Sn content of up to 2.5 wt.%, the compressive yield strength was slightly reduced. The finding might be attributed to the excessive liquid phase formed along with the visible appearance of particle boundary or grain coarsening. In conclusion, the study showed that the addition of Sn as sintering additive successfully promoted the Al alloys liquid phase sintering, especially with increased Sn content.

Acknowledgements

The authors would like to express sincere gratitude to the International Islamic University Malaysia (IIUM) and the Ministry of Higher Education (MOHE) Malaysia for sponsoring the current project under the grant number of FRGS/1/2019/TK08/UIAM/02/5.

Author details

Nur Ayuni Jamal[1*], Farazila Yusof[2], Yusilawati Ahmad[3], Norhuda Hidayah Nordin[1] and Suraya Sulaiman[4]

1 Manufacturing and Materials Engineering (MME) Department, Kulliyyah of Engineering, International Islamic University Malaysia (IIUM), Kuala Lumpur, Malaysia

2 Centre of Advanced Manufacturing and Material Processing (AMMP Centre), Universiti of Malaya, Kuala Lumpur, Malaysia

3 Biotechnology Engineering Department, Kulliyyah of Engineering, International Islamic University Malaysia (IIUM), Kuala Lumpur, Malaysia

4 Fakulti Teknologi Kejuruteraan Pembuatan dan Mekatronik, Universiti Malaysia Pahang (UMP), Pekan, Pahang, Malaysia

*Address all correspondence to: ayuni_jamal@iium.edu.my

IntechOpen

© 2021 The Author(s). Licensee IntechOpen. This chapter is distributed under the terms of the Creative Commons Attribution License (http://creativecommons.org/licenses/by/3.0), which permits unrestricted use, distribution, and reproduction in any medium, provided the original work is properly cited.

References

[1] Emamy M, Pourbahari B, Malekan M. Effects of Mg_2Sn intermetallic on the microstructure and tensile properties of Al–15% Mg_2Si–X% Sn composite. Journal of Materials Research. 2016;**31**(24):1-9. DOI: 10.1557/jmr.2016.426

[2] Banerjee S, Robi PS, Srinivasan A, Lakavath PK. Effect of trace additions of Sn on microstructure and mechanical properties of Al–Cu–Mg alloys. Materials and Design. 2010;**31**(8):4007-4015. DOI: 10.1016/j.matdes.2010.03.012

[3] Aliyu IK, Saheb N, Hassan SF, Al-Aqeeli N. Microstructure and properties of spark plasma sintered aluminum containing 1 wt.% SiC nanoparticles. Metals. 2015;**5**(1):70-83. DOI: 10.3390/met5010070

[4] Boland CD, Bishop D, Donaldson I, Hexemer Jr RL. On the development of an aluminum P/M Alloy for "Press-Sinter-Size" Technology. In: Proceedings of the International Conference on Powder Metallurgy & Particulate Materials (PowderMet 2010); 27-30 June 2010; Princeton: The Metal Powder Industries Federation (MPIF); 2010. pp. 1-14

[5] Qiu F, Gao X, Tang J, Gao YY, Shu SL, Han X, et al. Microstructures and tensile properties of Al–Cu matrix composites reinforced with nano-sized SiCp fabricated by semisolid stirring process. Metals. 2017;**7**(2):1-8

[6] Mahesh L, Reddy JS. Development and characterization of titanium nitride reinforced aluminium MMC's through powder metallurgy technique. Mechanics and Mechanical Engineering. 2017;**21**(1):29-36

[7] Ervina Efzan MN, Siti Syazwani N, Mohd Mustafa ABA. Fabrication method of aluminum matrix composite (AMCS): A review. Key Engineering Materials. 2016;**700**:102-110. DOI: 10.4028/www.scientific.net/KEM.700.102

[8] Ji X, Zhang C, Li S. In situ synthesis of core-shell-structured SiCp reinforcements in aluminium matrix composites by powder metallurgy. Metals. 2021;**11**(8):1-15. DOI: 10.3390/met11081201

[9] Feijoo I, Pena G, Cabeza M, Cristobal MJ, Rey P. MWCNT-reinforced AA7075 composites: Effect of reinforcement percentage on mechanical properties. Metals. 2021;**11**(6):1-19. DOI: 10.3390/met11060969

[10] Dolgopolov NA, Petelin AL, Rakov SV, Simanov AV. Penetration of liquid tin along grain boundaries and triple grain-boundary junctions of aluminum. Russian Journal of Non-Ferrous Metals. 2007;**48**(2):126-130. DOI: 10.3103/S1067821207020101

[11] Du X, Liu R, Xiong X, Liu H. Effects of sintering time on the microstructure and properties of an Al-Cu-Mg alloy. Journal of Materials Research and Technology. 2020;**9**(5):9657-9666. DOI: 10.1016/j.jmrt.2020.06.083

[12] Momeni H, Razavi H, Shabestari SG. Effect of supersolidus liquid phase sintering on the microstructure and densification of the Al-Cu-Mg pre-alloyed powder. Iranian Journal of Materials Science and Engineering. 2011;**8**(2):10-17

[13] Sercombe TB, Schaffer GB. On the use of trace additions of Sn to enhance sintered 2xxx series Al powder alloys. Materials Science and Engineering A. 1999;**68**(1-2):32-39. DOI: 10.1016/S0921-5093(99)00126-4

[14] Kondoh K, Kimura A, Takeda Y, Watanabe R. Behavior of magnesium

and tin at the surface of aluminum-silicon-magnesium-tin alloy powder particles at elevated temperature and their effect on direct nitriding reaction. Journal of the Japan Institute of Metals. 2000;**64**(11):1106-1112

[15] Azadbeh M, Razzaghi ZA. Properties evolution during transient liquid phase sintering of PM alumix 431. Advances in Materials Science and Engineering. 2009;**2009**(648906):1-5. DOI: 10.1155/2009/648906

[16] Liu ZY, Sercombe TB, Schaffer GB. The effect of particle shape on the sintering of aluminum. Metallurgical and Materials Transactions A. 2007;**38**:1351-1357. DOI: 10.1007/s11661-007-9153-2

[17] MacAskill IA, Hexemer RL Jr, Donaldson IW, Bishop DP. Effects of magnesium, tin and nitrogen on the sintering response of aluminum powder. Journal of Materials Processing Technology. 2010;**210**(15):2252-2260. DOI: 10.1016/j.jmatprotec.2010.08.018

[18] Schaffer GB, Huo SH, Drennan J, Auchterlonie GJ. The effect of trace elements on the sintering of an Al–Zn–Mg–Cu alloy. Acta Materilia. 2001;**49**(14):2671-2678. DOI: 10.1016/S1359-6454(01)00177-X

[19] Schaffer GB, Sercombe TB, Lumley RN. Liquid phase sintering of aluminium alloys. Materials Chemistry and Physics. 2001;**67**(1-3):85-91. DOI: 10.1016/S0254-0584(00)00424-7

[20] Suryanarayana C. Mechanical alloying and milling. Progress in Materials Science. 2001;**46**:1-184. DOI: 10.1016/S0079-6425(99)00010-9

[21] Pan Y, Lua X, Hayat MD, Yang F, Liu CC, Li Y, et al. Effect of Sn addition on the high-temperature oxidation behavior of high Nb containing TiAl alloys. Corrosion Science. 2020;**166**(108449):1-11. DOI: 10.1016/j.corsci.2020.108449

[22] Schaffer GB, Yao JY, Bonner SJ, Crossin E, Pas SJ, Hill AJ. The effect of tin and nitrogen on liquid phase sintering of Al–Cu–Mg–Si alloys. Acta Materilia. 2008;**56**(11):2615-2624. DOI: 10.1016/j.actamat.2008.01.047

[23] Aneta SN. The effect of the addition of boron on the densification, microstructure and properties of sintered 17-4 Ph stainless steel. Technical Transactions: Mechanics. 2014;**111**(2-M):85-96

[24] Ozay C, Gencer EB, Gokce A. Microstructural properties of sintered Al–Cu–Mg–Sn alloys. Journal of Thermal Analysis and Calorimetry. 2018;**134**:23-33. DOI: 10.1007/s10973-018-7171-5

[25] Gu ML, Zhang JH, Wei Z. Effect of sintering aid on microstructure and mechanical properties of TiB2/Tin tool materials. Advanced Materials Research. 2011;**287-290**:1933-1937. DOI: 10.4028/www.scientific.net/AMR.287-290.1933

[26] Padmavathi C, Upadhyaya A. Sintering behavior and mechanical properties of Al–Cu–Mg–Si–Sn aluminum alloy. Transactions of the Indian Institute of Metals. 2011;**64**:345-357. DOI: 10.1007/s12666-011-0089-2

[27] Massalski TB. Binary alloy phase diagrams. In: Okamoto H, Subramanian PR, Kacprzak L, editors. Advanced Materials. Ohio, USA: ASM International, Materials Park; 1991. p. 3589. DOI: 10.1002/adma.19910031215

[28] Harada Y, Jiang N, Kumai S. Mechanical properties of cold-rolled and annealed Al–12%Mg alloy sheet with high Mg solid solubility fabricated from vertical-type high-speed twin-roll cast strip. Materials Transactions. 2019;**60**(11):2435 2441. DOI: 10.2320/matertrans.f-m2019850

[29] Schaffer GB, Hall BJ. The influence of the atmosphere on the sintering of

aluminum. Metallurgical and Materials Transactions A. 2002;**33**(10):3279-3284. DOI: 10.1007/s11661-002-0314-z

[30] Dudas JH, Thompson CB. Improved sintering procedures for aluminum P/M parts. In: Hausner HH, editor. Modern Developments in Powder Metallurgy. Boston, MA: Springer; 1971. p. 490. DOI: 10.1007/978-1-4615-8963-1_2

[31] Atabay SE, Esen Z, Dericioglu AF. Effect of Sn alloying on the diffusion bonding behavior of Al-Mg-Si alloys. Metallurgical and Materials Transactions A. 2017;**48**:3181-3187. DOI: 10.1007/s11661-017-4089-7

Section 2

New Generation of Manufacturing Processes and Applications

Chapter 7

Drilling of 7075 Aluminum Alloys

Aishah Najiah Dahnel, Mohamad Noor Ikhwan Naiman, Muhammad Azim Mirza Mohd Farid, Ahmad Faris Abdul Rahman and Nur Munirah Meera Mydin

Abstract

Aluminum alloy (Al 7075) has been increasingly used as structural components in automotive and aerospace industry due to their low density, high strength and good corrosion resistance compared with other metals. To manufacture and assemble the components, drilling operations are often conducted. However, Al 7075 is ductile and soft, which causes difficulty in drilling, resulting in material adhesion, high tool wear, short tool life and poor hole quality. As a result of the poor hole quality, there is a high percentage of part rejection, which can increase the manufacturing time and cost. This chapter discusses challenges and techniques to drill Al 7075 in terms of the cutting parameters and drilling conditions to prolong the tool life and achieve good hole quality. Drilling experiments on Al 7075-T6 (heat-treated) were conducted using carbide cutting tools at various cutting parameters. Reducing cutting speed and increasing feed rate resulted in reducing tool wear, whereas a reduction in surface roughness, hence improved machined surface finish, was found when both cutting speed and feed rate were reduced in drilling Al 7075-T6. Producing good hole quality is vital during the drilling process to ensure a good assembly and product service performance.

Keywords: drilling, Al 7075, tool wear, hole quality, surface roughness, burr

1. Introduction

Aluminum alloys (particularly 6000 and 7000 grades) are desirable in various industrial applications due to their high strength to weight ratio and good corrosion resistance. Aluminum has a low density of 2.7 g/cm^3 compared to 7.87 g/cm^3 for steels. It is generally lighter (about one to three times) than steel; however, its properties (i.e., Al 7075) in terms of specific strength and toughness are almost similar to some steel. In addition, aluminum alloys are relatively cost-effective and have good economic value [1]. Some industries that highly rely on aluminum alloys are aerospace, automotive, shipbuilding and rail companies. In the industries, machining operations, particularly drilling, are widely conducted to manufacture the components. Drilling of the components is usually performed at the end of manufacturing processes to produce holes for assembly of a final product using suitable fasteners. Some defects such as cracks, burr and surface deformation may be introduced during drilling operations which could lead to poor product performance and product failure. It is important for the dimension of the drilled holes to be within the tolerance and has minimum or no defect, so that the components can be assembled securely as a functional product.

Drilling efficiency is strongly dependent on the materials, tool geometry, cutting parameters and process conditions. Drilling aluminum alloy in a dry condition is challenging as it often results in poor drilled hole quality, which includes the formation of burr, high surface roughness and adhesion on the work material. After the drilling operation, the product will need additional finishing processes, such as reaming and deburring to achieve the required hole quality. To overcome such issues, drilling can be done with the presence of cutting fluid to reduce heat generation and hence reduce thermal softening and material adhesion. Nevertheless, it is important to note that the use of cutting fluid or lubricant during drilling is costly, could have a high environmental impact and requires high maintenance for disposal. Therefore, dry drilling is usually preferable, however, choosing appropriate cutting parameters is crucial to produce drilled holes with good quality for the product assembly.

2. Application of aluminum alloy 7075

The increasing application of Al 7075 in the automotive and aerospace industries generally contributes to the reduction of the total mass of the vehicle. In the aerospace industry, Al 7075 is commonly used to produce structural components of the airframe, including wing skins, empennage or tail, and fuselage. The high corrosion resistance, high strength, high crashworthiness and durability of Al 7075 make the material desirable for structural components applications [2]. Typical mechanical properties for Al 7075 are shown in **Table 1**.

The modulus elasticity of AI 7075 is 71.7 GPa and its shear modulus is 26.9 GPa, which makes it suitable for high-performance metal application. Moreover, AI 7075 has a good tensile yield strength which means it could resist up to 503 MPa of tension before being deformed and will not revert to its initial form. The main reason for the high strength of Al 7075 compared to pure Aluminum is due to the zinc content as its primary alloying element [1]. The weight reduction while maintaining the strength of the component is advantageous to reduce fuel consumption and reduce maintenance costs. These good properties show the benefit and the reasons for Al 7075 is increasingly being used in industries for structural components.

From the aspect of machinability, Al 7075 is more machinable than other high-performance metals such as steel, cast iron and titanium due to its better chip formation and can be sheared easily [3]. Compared to other high-performance alloys (e.g., steel and titanium), the cutting forces, cutting temperature and energy consumed in machining aluminum alloy are relatively low which makes them a good alternative for achieving high productivity [4]. However, producing good hole quality in Al 7075 in terms of dimensional accuracy and good machines surface

Mechanical properties	Values
Ultimate tensile strength	572 MPa (83,000 psi)
Tensile yield strength	503 MPa (73,000 psi)
Shear strength	331 MPa (48,000 psi)
Fatigue strength	159 MPa (23,000 psi)
Modulus of elasticity	71.7 MPa (10,400 ksi)

Table 1.
Typical mechanical properties of Al 7075 tempered [1].

finish can be challenging. The soft and ductile properties of aluminum alloy could lead to thermal softening on the material during the operation which causes built-up edge formation that accumulated at the tool edge and material adhesion on the machined surface. Consequently, this leads to poor machined surface finish and hole quality. Hence, this chapter discusses the crucial drilling parameters that need to be considered to produce the required good hole quality.

3. Cutting tools

Cutting tools in terms of the tool geometry and tool material are important factors that need to be determined before conducting drilling operations because they can affect the tool wear, tool life and hence surface finish of the Al 7075. Twist drill, as shown in **Figure 1** (the image of the tool was captured by the authors), is the common type of drill that has been used, especially in drilling Al 7075 because it has helical grooves to facilitate the chip evacuation along with the flutes. The helix angle is the angle that forms between the leading edge of the drill and drill axis [5]. The efficiency of the drilling operation can be improved by using a drill with a low helix angle since the tool strength is improved by reducing the helix angle. For drilling Al 7075, the common helix angle that has been used is 30°. The point angle, which is the angle form between cutting lips is also important to ensure drilling efficiency. The standard point angle that is normally used for general purpose drilling is 118°.

The selection of drill material is important because it determines the toughness, hardness, wear-resistance and thermal resistance of the drill during the drilling process. The drill materials that have been used in drilling Al 7075 are commonly made from high-speed steels (HSS) and tungsten carbide (WC). It has been stated that the HSS drill is often used due to the availability in a wide range, low cost and hardness; however, the HSS drill does not resist high cutting temperature well [6]. To manufacture high-performance components, tungsten carbide drills are preferable as they usually produce better hole quality due to their ability to resist high cutting temperature and maintain their toughness compared to HSS drills. Less material adhesion, less BUE, low tool wear and improved chip formation were reported when using carbide drills during drilling Al 7075 compared to HSS drills [7]. This is supported by previous research [8] which found that tungsten carbide drill has 20% longer tool life than HSS drill, and it also produces high-quality holes with good dimensional accuracy within a range of 6.04 to 6.12 mm when drilling with 6 mm drill diameter in dry condition. Therefore, a tungsten carbide drill is generally recommended for drilling Al 7075.

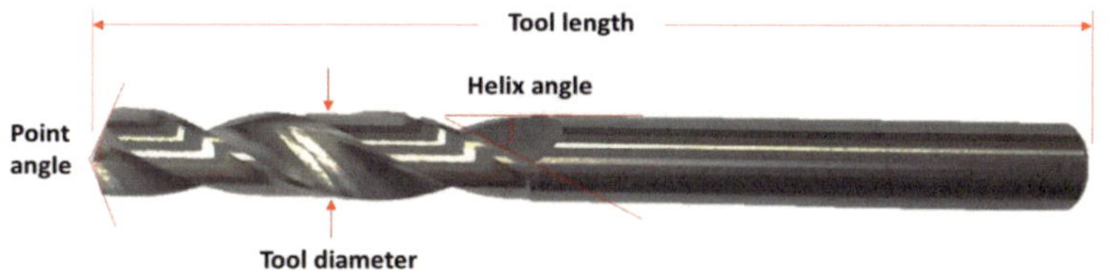

Figure 1.
Typical cutting tool for drilling aluminum alloys.

4. Cutting parameters

Cutting parameters are important in drilling operations because they have a significant influence in determining the rates of material removal, tool wear, and tool life which could affect the drilled hole quality. The important cutting parameters which need to be appropriately considered and selected in drilling operations are cutting speed and feed rate.

4.1 Cutting speed

A cutting speed is measured in terms of the rate at which the outside or periphery of the tool moves to the work being drilled. The range of cutting speeds that are normally used in drilling Al 7075 is within 50 to 250 m/min for 6 to 8 mm diameter cutting tools. It was reported that increasing cutting speed caused a reduction in the machined surface roughness due to the improved material shearing [9]. Based on [9], which conducted drilling experiments using 180, 200, 220, and 240 m/min of cutting speeds, it was found that the surface roughness of Al 7075 decreases by 5.49% when the cutting speed is increased from 180 to 240 m/min. However, an increase in cutting speeds could increase the tool chatter which affects the surface roughness of the machined surface. This view is supported by previous research [10], which also reported that increasing cutting speed could increase the tool vibration caused by the spindle rotation, which would lead to poor machined surface finish. To avoid the tool chatter and vibration due to high cutting speed during drilling, proper fixture and clamping of the work material, as well as a secured spindle head must be ensured before starting the drilling operation.

Furthermore, high cutting speed can also cause an increase in cutting temperatures between the tool and workpiece due to high heat generated during drilling operations which may lead to a higher tool wear rate. When the cutting speed is increased, the cutting temperature also increases which would result in the workpiece material sticking at cutting edges. This is supported by a previous study [11], which found that increasing cutting speed from 60 to 100 m/min caused the cutting temperature to increase from 195 to 240°C during drilling of Al 7075. Another previous study [9] shows that the flank wear increased from 0.08 to 0.19 mm due to an increase in cutting speeds from 180 to 240 m/min with 0.1 mm/rev feed rate which also resulted in decreased surface roughness from 4.015 to 3.619 μm. Although, higher cutting speed causes higher tool wear, using too low cutting speed is not recommended as it may cause the formation of Built-Up Edge (BUE) on the cutting edge which leads to high machined surface roughness and also cause low productivity. Based on [12], BUE formation was observed when drilling using a low cutting speed of 40 m/min with a cutting temperature of 160°C which resulted in surface roughness, Ra of 1.16 μm. Hence, using moderate cutting speeds within the range of 100 to 220 m/min is generally recommended when drilling Al 7075 to maintain good productivity.

4.2 Feed rate

Feed rate is a major factor that influences chip formation, cutting forces and hole quality. Feed rate is the distance that the drill moves into the workpiece for each complete turn of the cutting tool. The range of feed rates that are typically used in drilling operations of Al 7075 is within 0.01 to 0.10 mm/rev. A previous study [13] found that an increase in feed rate from 0.05 to 0.2 mm/rev at a constant cutting speed of 50 m/min caused the thrust force to increase from 825 to 1020 N and produce continuous chips which could entangle in the drill flutes hence causing poor machined surface finish. Another previous study [14] examined the effect of feed rate on drilled holes and found that at high feed rates, there is a deterioration in the shape of the holes. This is due to the high thrust forces caused by the thick chip formation. Generally, a low feed rate is recommended when drilling Al 7075 as it causes low thrust force, produces a good machined surface finish and prolongs the tool life.

5. Chip formation

Chips are formed during drilling operation as the aluminum alloy is removed by the cutting edges. Understanding the chip formation and morphology is important because it can influence the tool wear and hole quality. A drilling operation on Al 7075-T6 was conducted by the authors using 6.5 mm carbide drills at cutting speeds of 120 and 160 m/min as well as feed rates of 0.01 and 0.1 mm/rev. It was found that the chips produced are in the form of continuous, spiral, and discontinuous, as shown in **Figure 2a**.

The continuous chip in the form of a spiral or ribbon, as shown in **Figure 2b**, formed during drilling with increasing cutting speed. This is supported by previous research [15–17], which explained that the continuous chips formed due to material straining caused by thermal softening as a result of high cutting temperature. The material became more ductile, which led to the formation of continuous chips. Whereas, thinner continuous chips, as shown in **Figure 2c** are formed when the feed rate is decreasing. Continuous chips are typically undesirable because they could cause difficulty in chip evacuation as the chips entangle around the drill and jam in the flutes, which consequently could cause high surface roughness and poor drill hole quality [18]. In this case, a chip breaker is needed to break the chips into smaller pieces during drilling to facilitate chip evacuation from the drilled hole.

Based on [18], the chip becomes shorter as the drilling operation progresses as the resistance between the material and the drilling bit decreases. In addition, the chip produced unfolds due to the opposite of the torque of the friction to the chip rotation. The break-up of the chip into pieces is when the chip reaches its fracture point, whereas the continuous cone-shaped spiral chips break into smaller pieces when the chip exceeds the breaking torque between the tool and hole wall [18]. In addition, the use of chilled air during drilling Al 7075 with carbide drill could be favorable as it was found to result in shorter chips than dry drilling by 20 and 32.5% at cutting speeds of 120 and 160 m/min, respectively, as shown in **Figure 3**. The chip started to change into a segmented chip from a continuous chip due to the lower cutting temperature which lowered the thermal softening and material straining.

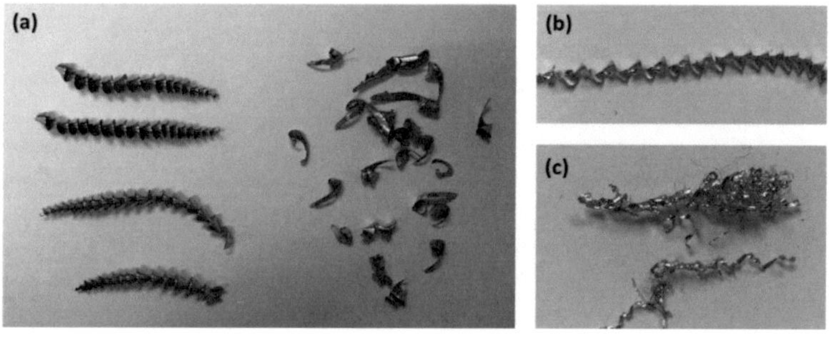

Figure 2.
Chip morphology of Al 7075-T6 during drilling operations (a) continuous, discontinuous and smaller pieces chips (b) cone-shaped spiral chips (c) thinner continuous chips.

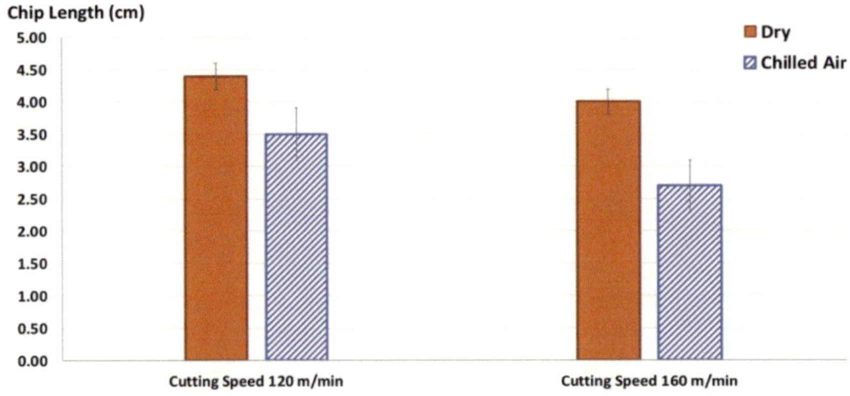

Figure 3.
Comparison of chip length when drilling Al 7075-T6 at cutting speeds of 120 and 160 m/min and a constant feed rate of 0.05 mm/rev.

6. Tool wear

Tool wear is an important element that can affect tool life, the machined surface finish of the drilled hole and hence hole quality. Tool wear is the change in the geometry and dimension, particularly at the cutting edges, which indicates the failure rate of the cutting tool due to drilling a significant number of holes. The tool wear mechanisms that typically occur in drilling Al 7075 are abrasive and adhesive wear, as shown in **Figure 4**, which was observed during experiments conducted by the authors. The abrasive wear, as shown in **Figure 4a**, occurs as the tool material (e.g., carbide and cobalt), especially at the cutting edges, were abraded due to the removing process of the harder workpiece material (e.g., Al7075) in the drilling process.

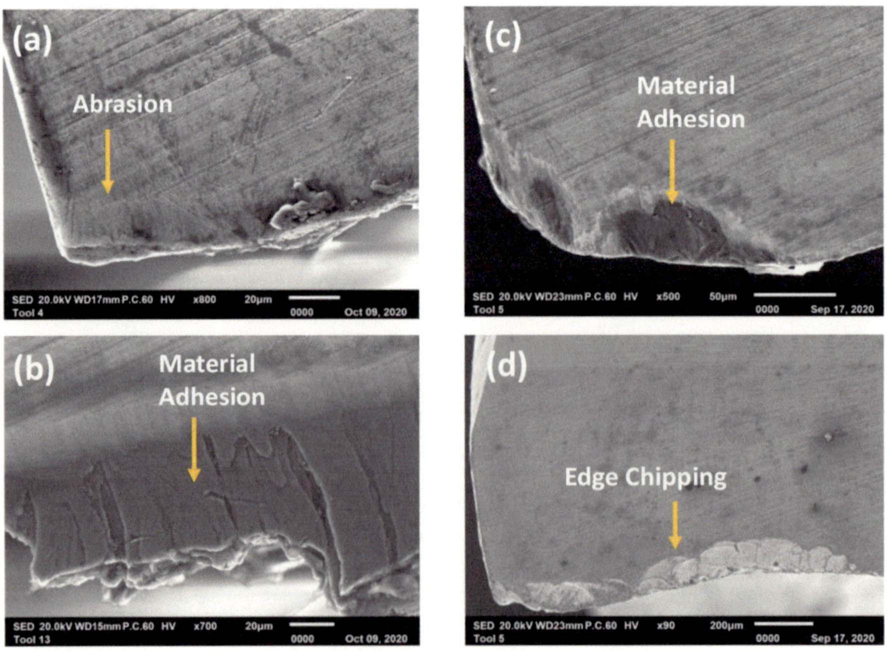

Figure 4.
Tool wear mechanism when drilling Al 7075-T6 (a) abrasive wear, (b, c) adhesion of material, and (d) chipping at the cutting edge of carbide drills.

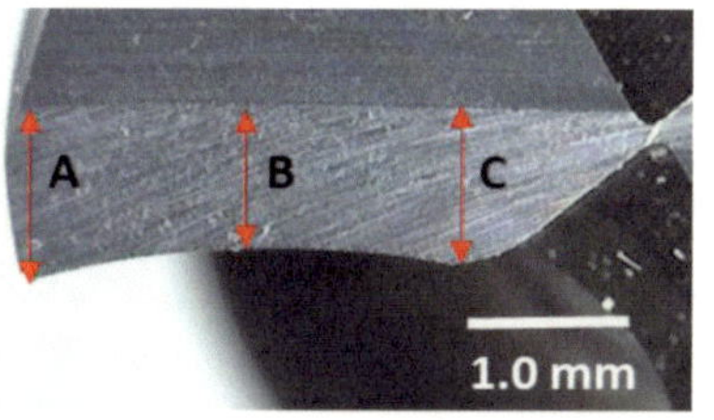

Figure 5.
Measurement of tool wear is taken as the change in flank length indicated by A, B, and C after drilling each 10th hole compared to the initial flank length.

Based on [19], scratches at the cutting edges parallel to the cutting direction will show the proof and region of abrasive wear. It was reported [20] that the abrasion mechanism occurs at a low cutting speed due to a low chip removal rate compared to drilling at a higher cutting speed. To minimize the abrasive wear, a cutting tool with a harder material than the workpiece material should be used in the drilling operation. Adhesive wear, as shown in **Figure 4b** and **c**, occurs when the chip or machined material adhere to the cutting edges due to high cutting temperature and pressure, which can cause edge chipping, as can be seen in **Figure 4d** when the adhered material breaks off [21, 22]. Previous studies [20, 23] reported that increasing cutting speed from 76 to 198 m/min in dry drilling resulted in material adhesion due to an increase in cutting temperature, thus will lead to higher tool wear.

The tool wear is typically affected by the cutting parameters and process conditions used. The authors experimented drilling Al 7075-T6 heat-treated using 6.5 mm carbide drills to investigate the effect of cutting speed and feed rate on tool wear. The tool wear was determined by measuring the change in the flank, as shown in **Figure 5**. This measurement follows ISO 3685:1993, which recommends that the end of tool life is reached when the flank wear is equivalent to 0.3 mm.

From the experiment conducted by the authors, it can be seen in **Figure 6** that increasing cutting speed from 80 to 120 and to 160 m/min at a constant feed rate of 0.05 mm/rev resulted in increasing tool wear. This is likely due to increasing heat generation, which caused weakening of the tool material as well as material adhesion on the cutting edges. Feed rate also has a significant effect on tool wear. As shown in **Figure 7**, increasing feed rate from 0.01 to 0.05 and to 0.1 mm/rev

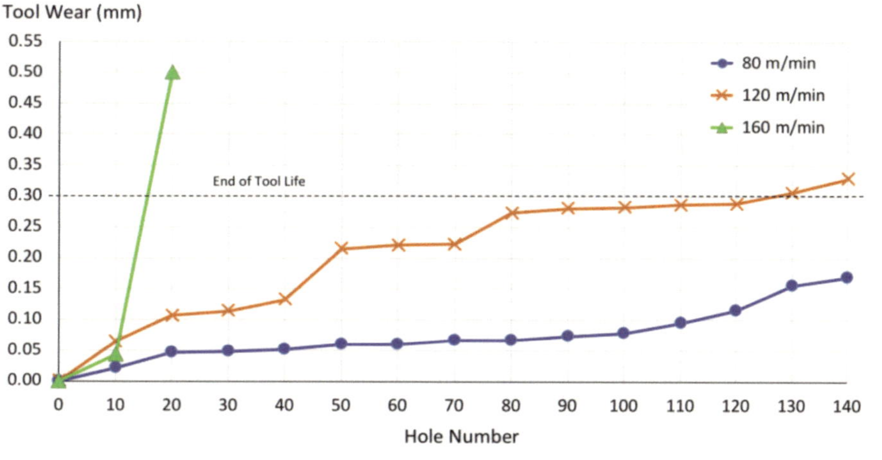

Figure 6.
Tool wear in drilling Al 7075-T6 (heat-treated) using 6.5 mm carbide drills at cutting speeds of 80, 120 and 160 m/min and a constant feed rate of 0.05 mm/rev.

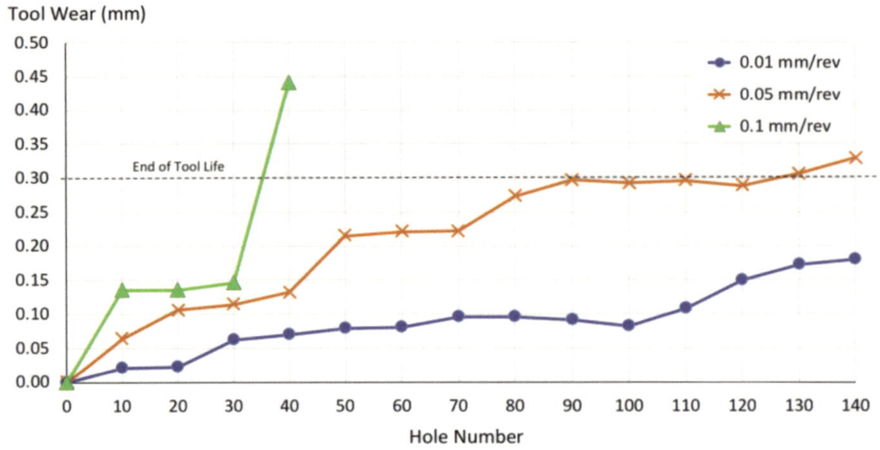

Figure 7.
Tool wear in drilling Al 7075-T6 (heat-treated) using 6.5 mm carbide drills at feed rates of 0.01, 0.05 and 0.1 mm/rev and a constant cutting speed of 120 m/min.

at a constant cutting speed of 120 m/min was found to increase tool wear. This is likely due to the increasing volume of material removed per tool rotation that could cause higher thrust force and weaken the cutting edges leading to higher tool wear. This indicates that a low cutting speed and low feed rate can produce low tool wear therefore a longer tool life.

Even though using a low cutting speed and a low feed rate can cause a lower tool wear rate in drilling Al 7075-T6, it is usually not preferable in the industry as this can lead to longer production time hence low productivity. Therefore, moderate cutting speed and feed rate could be used when productivity is a concern, although frequent tool change may be needed to ensure the tool wear does not affect the hole quality. Drilling in a dry condition typically results in high tool wear due to heat generated between workpiece material and cutting tool. Therefore, the presence of cutting fluid and chilled air during drilling Al 7075 could improve the tool wear. Referring to [24], chilled air can act as a coolant to reduce the temperature at the cutting zone, thus will result in low tool wear, hence longer tool life.

7. Hole quality

Ensuring that the holes are produced with good quality as required is important for component assembly hence the product functionality and service life. The drilled hole quality can be measured and assessed in terms of the hole diameter, roundness, machined surface roughness and burr formation.

7.1 Hole diameter and roundness

The diameter of drilled holes has been reported to increase when the cutting speed, feed rate and point angle increase [25]. The increase in cutting speed and feed rate leads to the increase in the oversized hole because of the vibrations that are induced at higher speed and feed rate [10, 26]. HSS drills have been reported to produce undersized holes compared to carbide drills [26]. Therefore, carbide drills are recommended for drilling the Al alloy to produce holes within the required tolerance. The authors conducted a drilling experiment on Al 7075-T6 (thickness of 13 mm) using 6.5 mm diameter carbide tools with two different feed rates of 0.01

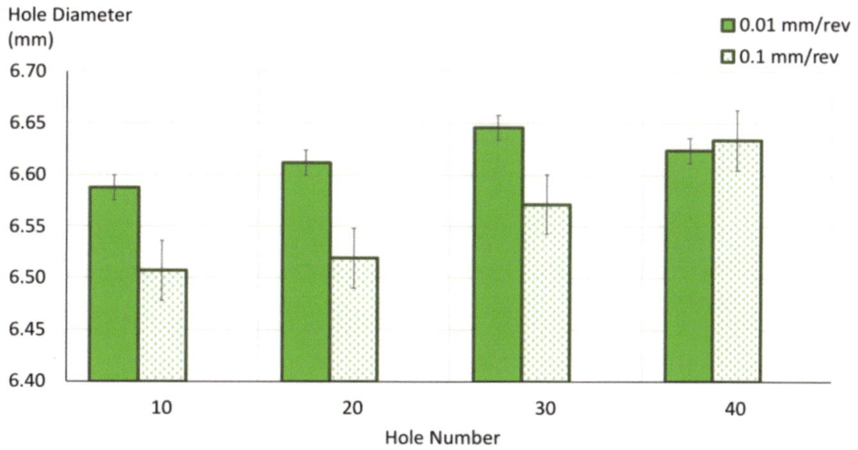

Figure 8.
The effect of feed rate on hole diameter in drilling Al 7075-T6 (heat-treated) at a constant cutting speed of 120 m/min and 6.5 mm diameter carbide cutting tools.

and 0.1 mm/rev at a constant cutting speed of 120 m/min. The drilled hole diameters were measured and results are shown in **Figure 8**. Increasing hole numbers resulted in increasing hole diameter, which is likely caused by the increased chip adhesion on the cutting tool. Generally, the use of a higher feed rate of 0.1 mm/rev resulted in more accurate hole diameters closer to the tool nominal diameter of 6.5 mm.

In addition, hole roundness is also an aspect that determines the drilled hole quality. The roundness of the hole is defined as how closely the hole matches a perfect circle, which can be determined by measuring the consistency of the hole diameters at various orientations. The hole roundness error has been reported to be significantly influenced by the feed rate [15]. Based on [15], which conducted experiments on drilling aluminum alloy using a 6 mm diameter drill, the hole roundness error was found to be lower (0.030–0.038 mm) when using a lower feed rate of 0.04 mm/rev compared to the higher feed rate of 0.08 mm/rev which causes higher roundness error (0.05–0.06 mm). However, no significant difference in the roundness error was observed when cutting speeds changed from 25 to 50 m/min [15]. Therefore, using a low feed rate is recommended to ensure the hole is produced with minimum dimensional error. However, as cutting speed is likely to have less effect on the hole diameter, a higher cutting speed is recommended to improve productivity. Nevertheless, the tool wear needs to be observed as the material adhesion could also affect the hole dimensional accuracy.

7.2 Surface roughness

The machined surface typically contains irregularities and deviations from the desired form due to the machining operations, cutting parameters, and cutting conditions used. These deviations are normally assessed as roughness. The surface roughness acts as an indicator to determine the quality of the holes and surface finish. Cutting parameters are seen to be the most critical factor that could affect the surface roughness. The authors conducted a drilling experiment on Al7075-T6 in two different conditions (dry and with chilled air) and different cutting parameters to determine their effects on the machined surface roughness (Ra). The pressure and flow rate of the chilled air that was used during the drilling were 14 m/s and 6.9 bar, respectively. The surface roughness results, which were measured in terms of Ra are shown in **Figure 9**.

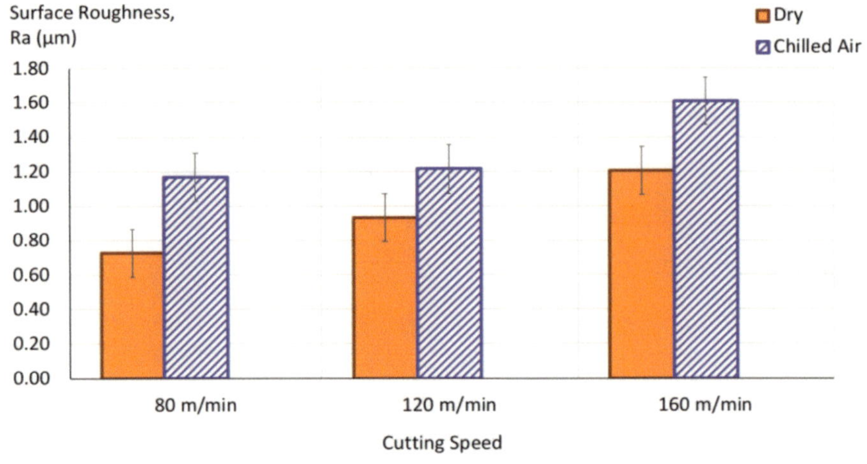

Figure 9.
The effect of cutting speed and condition on surface roughness (Ra) in drilling Al 7075-T6 (heat-treated) using cutting speeds of 80, 120 and 160 m/min at a constant feed rate of 0.05 mm/rev and 6.5 mm diameter carbide drills.

Increasing cutting speed from 80 to 120 and to 160 m/min resulted in increasing surface roughness by the values as shown in **Figure 9**. This is supported by a previous study [12], which also conducted drilling of Al 7075 alloy with three parameters which are cutting speeds (within range of 40 to 120 m/min), feed rates (within range of 0.05 to 0.15 mm/rev), and point angle (within range of 120 to 140°). It was found that an increase in cutting speeds from 40 to 120 m/min at a constant feed rate of 0.05 mm/rev and point angle of 140° resulted in increased surface roughness value, Ra from 0.5 to 1.16 µm. During drilling operation at higher cutting speeds, the ductility properties of aluminum alloy will increase because of the higher heat generated that is caused by, the higher spindle speed and energy. The higher cutting temperature causes thermal

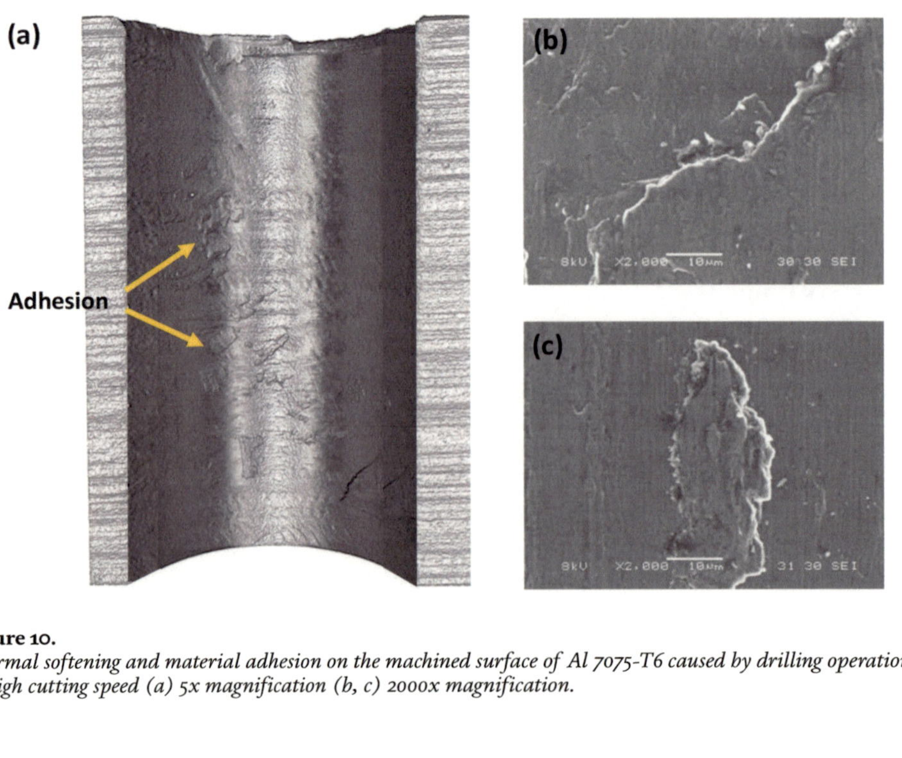

Figure 10.
Thermal softening and material adhesion on the machined surface of Al 7075-T6 caused by drilling operations at high cutting speed (a) 5x magnification (b, c) 2000x magnification.

softening of the workpiece, consequently, more material adhesion on the machined surface, as shown in **Figure 10**, hence higher surface roughness (Ra).

Furthermore, the increasing feed rate when drilling Al 7075-T6 was also found to result in increased surface roughness, as shown in **Figure 11**. The material removal rate increases when the feed rate increases, which leads to higher cutting forces hence poor surface finish of drilled holes and higher Ra value. Whereas, using chilled air (10°C) during drilling resulted in 25–60% higher Ra than dry drilling, as can be seen in **Figures 6** and **8**. Therefore, using chilled air in drilling Al 7075 is not favorable in terms of surface finish. This is likely due to chilled air causing which works hardening of the chip, which causes poor machined surface finish as the chip evacuates from the holes. The higher Ra when using a higher feed rate is typically due to the feed mark on the machined surface as shown in **Figure 12**. The application of high-pressure internal water-based cutting fluid has been recommended in drilling aluminum alloy to result in lower surface roughness [27]. The assistance of high-pressure cutting fluid (i.e., with a pressure of 50 bar) can improve chip evacuation hence less tendency of the chip scratching the machined surface. However,

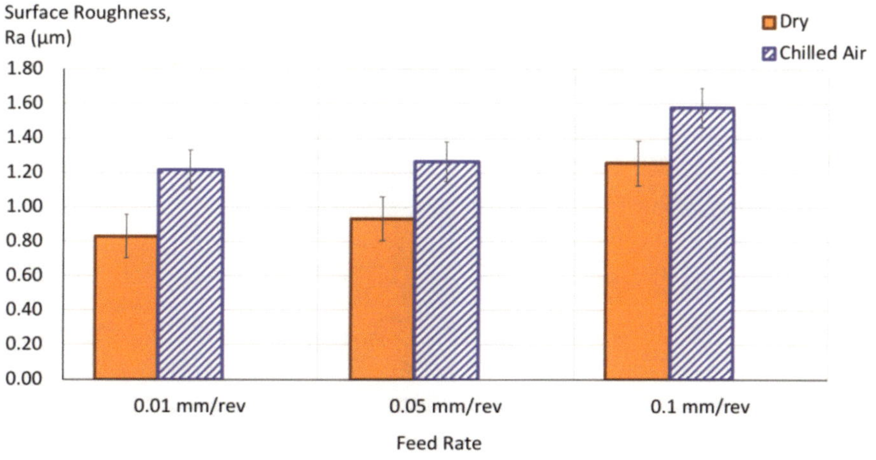

Figure 11.
The effect of feed rate and condition on surface roughness (Ra) in drilling Al 7075-T6 (heat-treated) using feed rates of 0.01, 0.05 and 0.1 mm/rev at a constant cutting speed of 120 m/min and 6 mm diameter carbide drills.

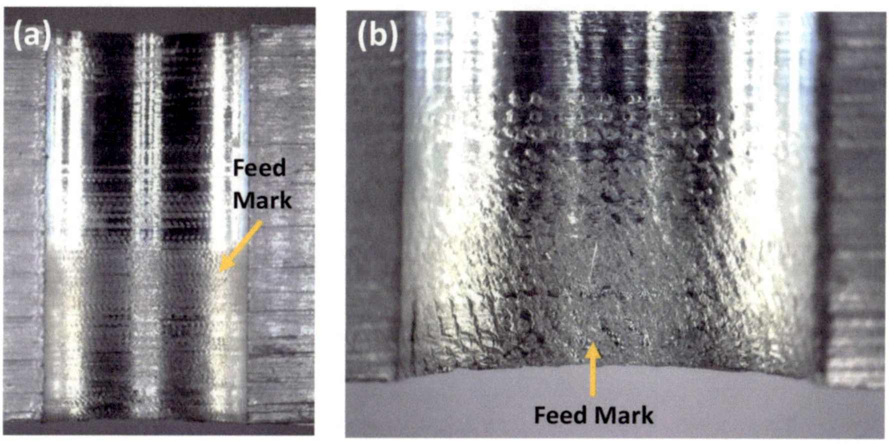

Figure 12.
Feed mark on the machined surface of Al 7075-T6 (heat-treated) caused by drilling at a high feed rate of 0.1 mm/rev; (a) the surface of the drilled hole that was sectioned into a half (b) the hole exit.

careful handling and appropriate fluid disposal must be practiced to avoid environmental pollution and health issues to operators.

7.3 Burr formation

Burrs are the common defect that occurred due to the machining process. A burr is the protrusion of workpiece material typically at the hole edges (entry and exit of the hole), as shown in **Figure 13a** and **b**. The burr height is measured as indicated in **Figure 13b**. The burr is hard and sharp, which can cause difficulty in assembly and may cause injury to the operators. Therefore, the burr is unwanted and needs to be deburred to ensure the good assembly of the parts.

The machining process could result in primary and secondary burrs. The primary burr occurs during the drilling operations after the material has been removed by the cutting edges. The secondary burr is the remaining material at the edge of the drilled hole after breakage of the primary burr due to deburring process. The formation of burr depends on the cutting parameters (cutting speed and feed rate), ductility of workpiece material, and tool geometry especially point angle. The authors experimented on dry drilling of Al 7075-T6 using carbide drills and it was found that using a higher cutting speed of 120 m/min generally resulted in a higher burr compared to 80 m/min, as shown in **Figure 14**. This is likely due to an increase

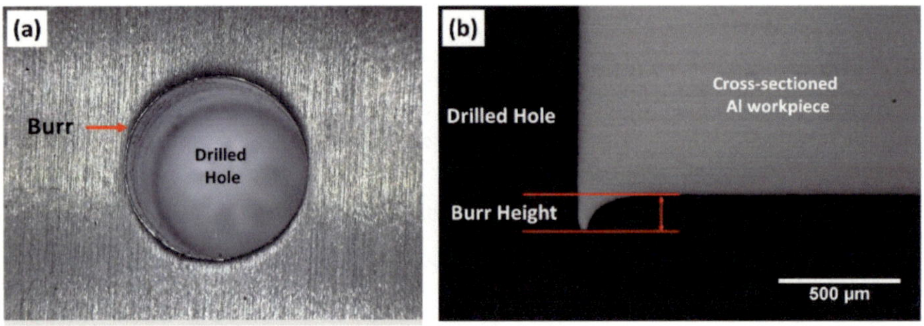

Figure 13.
(a) Burr at the edge of the drilled hole (b) Burr height which is measured after the hole was sectioned into a half.

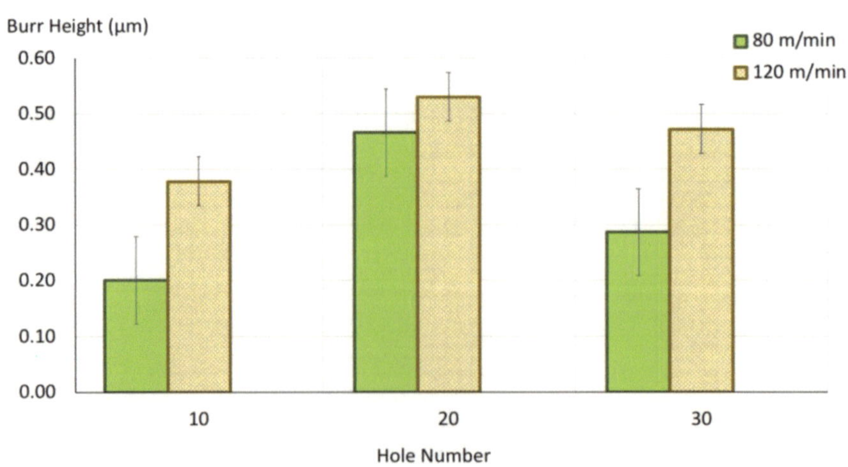

Figure 14.
The burr height when drilling Al 7075-T6 (heat-treated) using cutting speeds of 80 and 120 m/min at a constant feed rate of 0.05 mm/rev.

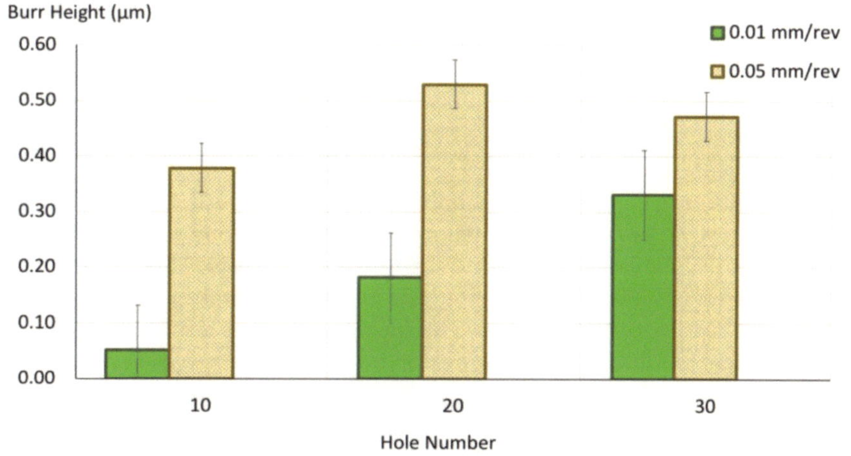

Figure 15.
The burr height when drilling Al 7075-T6 (heat-treated) using feed rates of 0.01 and 0.05 mm/rev at a constant cutting speed of 120 m/min.

in ductility of the alloy as more heat is generated with a higher cutting speed, causing the material to project at the edge [28]. In addition, the burr formation can also be influenced by feed rate. As shown in **Figure 15**, the usage of a higher feed rate of 0.05 mm/rev resulted in higher burr compared to the lower feed rate of 0.01 mm/rev. This is likely due to the higher volume of material removed.

A previous study [7] on the burr formation of Al 5083 caused by cutting parameters has also been conducted, in which the finding could be related to the case of Al 7075. Feed rates within the range of 0.04 to 0.14 mm/rev and cutting speed range within 19 to 57 mm/min were used [7]. It was found that the feed rate highly influences the growth of burrs during drilling compared to cutting speed. Drilling at feed rate 0.14 mm/rev with 57 m/min cutting speed produced high burr formation, which is 7 μm, while drilling at 0.04 mm/rev feed rate produced the lowest burr formation which is 3.8 μm. The burrs resulting from drilling Al 5083 were observed to be more visible at the exit holes compared to entry holes. Therefore, using a low cutting speed and feed rate in drilling Al 7075 could be beneficial to minimize the burr formation, however, this may cause low productivity in the industry.

8. Future development in drilling technology

Drilling operations have always been necessary for the manufacturing and assembly of mechanical components, which are made of aluminum alloys. This motivates the industry and researchers to further investigate the drilling technology that can produce good hole quality with longer tool life, hence optimizing productivity. Ultrasonic Assisted Drilling (UAD), which employs a cutting tool that is vibrated during the material removal process, has been receiving attention in the industry. The application of UAD on high-performance materials, including aluminum alloys, Carbon Fiber Reinforced Polymer (CFRP), and titanium alloys, has shown potential to reduce tool wear, reduce thrust forces and increase material removal rate, which is reported to be due to intermittent cutting and reduced chip resistance [21, 29]. In addition, the use of cryogenic coolant (i.e., carbon dioxide and liquid nitrogen) could be useful to facilitate heat removal during the drilling process, especially for deep hole production, which could result in less material adhesion and improved hole surface finish. Therefore, further research involving these drilling technologies, particularly on aluminum alloys, is necessary for industrial applications.

9. Conclusions

In conclusion, tool wear and hole quality of drilled holes of Al 7075 are highly influenced by cutting parameters, cutting tool and conditions during the drilling operations. Results showed that lower cutting speed and lower feed rate are favorable to produce low tool wear and better hole quality in terms of diameter, roundness, surface roughness and burr. Low cutting speed causes less heat generation, hence there is less tendency of chip adhesion on the cutting edges and machined surfaces. This can result in low surface roughness as well as less bur formation, therefore leading to good hole quality. Whereas, low feed rate means there is less material removed per one rotation of the tool, which results in a smaller chip, better chip evacuation, low cutting forces and hence good machined surface finish. However, it is important to note that when using lower cutting speed and lower feed rate, the workpiece cutting length per minute is slower, and this leads to longer time consumption in producing the hole. To improve productivity, higher cutting speed and feed rate may be used in drilling Al 7075, however, cutting fluid should be used to dissipate the heat and facilitate chip evacuation hence producing good hole quality. Nevertheless, careful consideration needs to be taken in handling and disposing of the fluid to avoid environmental pollution and also health issues to the operators.

Acknowledgements

This publication is supported and funded by the Ministry of Higher Education Malaysia and International Islamic University Malaysia (IIUM) under Fundamental Research Grant Scheme, FRGS/1/2018/TK03/UIAM/03/5.

Conflict of interest

The authors declare no conflict of interest.

Author details

Aishah Najiah Dahnel*, Mohamad Noor Ikhwan Naiman,
Muhammad Azim Mirza Mohd Farid, Ahmad Faris Abdul Rahman
and Nur Munirah Meera Mydin
Department of Manufacturing and Materials Engineering, International Islamic
University Malaysia, Kuala Lumpur, Malaysia

*Address all correspondence to: aishahnajiah@iium.edu.my

IntechOpen

© 2022 The Author(s). Licensee IntechOpen. This chapter is distributed under the terms of the Creative Commons Attribution License (http://creativecommons.org/licenses/by/3.0), which permits unrestricted use, distribution, and reproduction in any medium, provided the original work is properly cited.

References

[1] Cavallo C. All about 7075 Aluminum (Properties, Strength, and Uses). 2021. Thomas Publishing Company. Available from: https://www.thomasnet.com/articles/metals-metal-products/all-about-7075-aluminum-properties-strength-and-uses/ [Assessed: November 30, 2021]

[2] Hirsch J. Recent development in aluminium for automotive applications. Transactions of Nonferrous Metals Society of China. 2014;**24**(7):1995-2002. DOI: 10.1016/S1003-6326(14)63305-7

[3] Santos MC, Machado AR, Sales WF, Barrozo MA, Ezugwu EO. Machining of aluminum alloys: A review. The International Journal of Advanced Manufacturing Technology. 2016;**86**(9): 3067-3080. DOI: 10.1007/s00170-016-8431-9

[4] Schuster PA, Österreicher JA, Kirov G, Sommitsch C, Kessler O, Mukeli E. Characterisation and comparison of process chains for producing automotive structural parts from 7xxx aluminium sheets. Metals. 2019;**9**(3):305. DOI: 10.3390/met9030305

[5] Ren K, Ni J. Analyses of drill flute and cutting angles. The International Journal of Advanced Manufacturing Technology. 1999;**15**(8):546-553. DOI: 10.1007/s001700050100

[6] Liu D, Tang Y, Cong WL. A review of mechanical drilling for composite laminates. Composite Structures. 2012;**94**(4):1265-1279. DOI: 10.1016/j.compstruct.2011.11.024

[7] Aamir M, Giasin K, Tolouei-Rad M, Vafadar A. A review: Drilling performance and hole quality of aluminium alloys for aerospace applications. Journal of Materials Research and Technology. 2020;**9**(6): 12484-12500. DOI: 10.1016/j.jmrt.2020.09.003

[8] Nouari M, List G, Girot FA, Gehin D. Effect of machining parameters and coating on wear mechanisms in dry drilling of aluminium alloys. International Journal of Machine Tools and Manufacture. 2005;**45**(12-13):1436-1442. DOI: 10.1016/j.ijmachtools.2005.01.026

[9] Bhushan RK, Kumar S, Das S. Effect of machining parameters on surface roughness and tool wear for 7075 Al alloy SiC composite. The International Journal of Advanced Manufacturing Technology. 2010;**50**(5-8):459-469. DOI: 10.1007/s00170-010-2529-2

[10] Sultan AZ, Sharif S, Kurniawan D. Effect of machining parameters on tool wear and hole quality of AISI 316L stainless steel in conventional drilling. Procedia Manufacturing. 2015;**2**:202-207. DOI: 10.1016/j.promfg.2015.07.035

[11] Bagci E, Ozcelik B. Analysis of temperature changes on the twist drill under different drilling conditions based on Taguchi method during dry drilling of Al 7075-T651. The International Journal of Advanced Manufacturing Technology. 2006;**29**:629-636. DOI: 10.1007/s00170-005-2569-1

[12] Yaşar N, Boy M, Günay M. The effect of drilling parameters for surface roughness in drilling of AA7075 alloy. MATEC Web of Conferences. 2017;**112**:01018. DOI: 10.1051/matecconf/201711201018

[13] Salem SB, Bayraktar E, Boujelbene M, Katundi D. Effect of cutting parameters on chip formation in orthogonal cutting. Journal of Achievements in Materials and Manufacturing Engineering. 2012;**50**(1):7-17. DOI: 10.1063/1.3552327

[14] Kurt MU, Kaynak YU, Bagci E. Evaluation of drilled hole quality in Al 2024 alloy. The International Journal of Advanced Manufacturing Technology.

2008;**37**(11):1051-1060. DOI: 10.1007/s00170-007-1049

[15] Uddin M, Basak A, Pramanik A, Singh S, Krolczyk GM, Prakash C. Evaluating hole quality in drilling of Al 6061 alloys. Materials. 2018;**11**(12):2443. DOI: 10.3390/ma11122443

[16] Xu D, Feng P, Li W, Ma Y, Liu B. Research on chip formation parameters of aluminum alloy 6061-T6 based on high-speed orthogonal cutting model. The International Journal of Advanced Manufacturing Technology. 2014;**72**(5):955-962. DOI: 10.1007/s00170-014-5700-3

[17] Lorentzon J, Järvstråt N, Josefson BL. Modelling chip formation of alloy 718. Journal of Materials Processing Technology. 2009;**209**(10):4645-4653. DOI: 10.1016/j.jmatprotec.2008.11.029

[18] Zheng L, Wang C, Yang L, Song Y, Fu L. Characteristics of chip formation in the micro-drilling of multi-material sheets. International Journal of Machine Tools and Manufacture. 2012;**52**(1):40-49. DOI: 10.1016/j.ijmachtools.2011.08.017

[19] Diniz AE, Machado ÁR, Corrêa JG. Tool wear mechanisms in the machining of steels and stainless steels. The International Journal of Advanced Manufacturing Technology. 2016;**87**(9):3157-3168. DOI: 10.1007/s00170-016-8704-3

[20] Arsecularatne JA, Zhang LC, Montross C. Wear and tool life of tungsten carbide, PCBN and PCD cutting tools. International Journal of Machine Tools and Manufacture. 2006;**46**(5):482-491. DOI: 10.1016/j.ijmachtools.2005.07.015

[21] Dahnel AN, Ascroft H, Barnes S. The effect of varying cutting speeds on tool wear during conventional and ultrasonic assisted drilling (UAD) of carbon fibre composite (CFC) and titanium alloy stacks. Procedia Cirp. 2016;**46**:420-423. DOI: 10.1016/j.procir.2016.04.044

[22] Khatri A, Jahan MP. Investigating tool wear mechanisms in machining of Ti-6Al-4V in flood coolant, dry and MQL conditions. Procedia Manufacturing. 2018;**26**:434-445. DOI: 10.1016/j.promfg

[23] Jindal A. Analysis of tool wear rate in drilling operation using scanning electron microscope (SEM). Journal of Minerals and Materials Characterization and Engineering. 2012;**11**(01):43. DOI: 10.4236/jmmce.2012.111004

[24] Rubio EM, Agustina B, Marín M, Bericua A. Cooling systems based on cold compressed air: A review of the applications in machining processes. Procedia Engineering. 2015;**132**:413-418. DOI: 10.1016/j.proeng.2015.12.513

[25] Uzun G, G. U. Effect of cutting parameters on the drilling of AlSi7 metallic foams. Materiali in Technologije (Materials and Technology). 2017;**51**(1):19-24. DOI: 10.17222/mit.2015.106

[26] Zhang PF, Churi NJ, Pei ZJ, Treadwell C. Mechanical drilling processes for titanium alloys: A literature review. Machining Science and Technology. 2008;**12**(4):417-444. DOI: 10.1080/10910340802519379

[27] Popan IA, Popan AI, Carean A, Fratila D, Trif A. Study on chip fragmentation and hole quality in drilling of aluminium 6061 alloy with high pressure internal cooling. MATEC Web of Conferences. 2019;**299**:04014. DOI: 10.1051/matecconf/201929904014

[28] Mydin NM, Dahnel AN, Raof NA, Khairussaleh NK, Mokhtar S. The effect of chilled air on Burr formation when drilling Aluminium alloy in manufacturing industry. International Journal of Progressive Sciences and Technologies. 2021;**28**(1):437-445

[29] Chu NH, Nguyen VD, Ngo QH. Machinability enhancements of ultrasonic-assisted deep drilling of aluminum alloys. Machining Science and Technology. 2020;**24**(1):112-135. DOI: 10.1080/10910344.2019.1636267

Chapter 8

Application of the Aluminothermic Reduction Process for Magnesium Removal in Aluminum Scrap

Rocio Maricela Ochoa Palacios, Citlaly Castillo Rodriguez, Jesus Torres Torres, Perla Janet Resendiz Hernandez and Alfredo Flores Valdes

Abstract

Magnesium is considered as impurity element in aluminum recycled for obtaining some cast alloys, with low concentration Mg, because at 0.1 wt% results in fragility, fractures, and defects. This research applies the aluminothermic reduction process to decrease magnesium content in aluminum cans by adding ZnO, to produce reaction products solid-state (Al_2O_3, MgO and $MgAl_2O_4$), and there is a possibility to obtain Al-Zn alloy. The conditions of the process were, melting temperature (750, 800, 850°C) and stirring velocity (200, 250, 300 rpm). The Mg and Zn contents were measured for chemical analysis and scrap generated from every process was analyzed by X-ray diffraction. The results show how the aluminothermic reduction decreased Mg from 0.93 to 0.06 wt% and increased zinc up to 5.52wt % in the molten metal. Therefore, this process can be used to remove Mg and can also prevent the generation of polluting gases into the environment.

Keywords: aluminum, scrap, aluminothermic, magnesium removal, Al-Zn alloy

1. Introduction

Aluminum alloys are a material required mainly in aerospace and automotive sectors [1]. This has contributed to an increased demand for aluminum and therefore the search for new forms of production. Recycling products have been considered as an alternative source of raw material; however, the secondary aluminum industry is facing up the implementation of different processes to return these materials to everyday life, modifying the mechanical properties according to the type of application to which it will be destined. For this reason, the trend toward metal recycling has increased significantly in recent years, not only because of economic aspects, but also because of the positive impacts on the environment [2–4]. Nevertheless, there is still a lot of development to be done because the

IntechOpen

industry demands high-quality materials regardless of their origin. For cast alloys, the aluminum scrap must be carefully selected because the impurities degrade the material form that decrease the mechanical properties of the alloys. The most common impurities in recycled alloys are Fe, Si, and Mg in the automotive industry [5, 6]. Different refining processes are used for the removal of magnesium, such as Cl_2 or Fl_2 gas injection, but the processes produce great amounts of HCl gas. Many countries have been forced to regulate this process, while others have banned it due to the high amount of polluting gases that are emitted into the environment [7–9]. Right to this, the industry needs to find new technologies to reduce the magnesium in aluminum scraps, such as the process of compound separation methods or aluminothermic reduction with different oxides [10–13].

This work uses the aluminothermic reduction process with ZnO powder, with the objective of removing the magnesium content in aluminum molten scrap and adding zinc for production aluminum alloys type 700. By implementing this technology, gases can be considerably reduced because the reactions produced in the metal bath can lead to the formation of magnesium products in a solid-state and to be easily removed with the slag.

2. Experimental procedure

The study used aluminum scrap from beverage cans as raw material in the aluminothermic reduction process with ZnO powder as reactive, the composition of both materials were analyzed by atomic emission spectroscopy (AES), the equipment used was SPECTRO model M11, the results are shown in **Table 1**. The beverage cans were melted in 10 kg capacity induction furnace, then, the melted aluminum was degassed using N_2 gas for 10 minutes to obtain secondary aluminum ingots cooled at room temperature. These ingots were cut into pieces of 2 kg to be melted again in a laboratory-scale induction furnace and carried out aluminothermic reduction process using ZnO powder at 750, 800, and 850°C and 200, 250, and 300 rpm of stirring velocities. **Figure 1** shows the induction furnace where the aluminothermic reduction was carried out, (a) top view photography of the furnace, (b) scheme of the interior furnace, and (c) the impeller used during the experimentation. The time of all experiments was given for 60 minutes, taking samples every 10 minutes to measure the magnesium and zinc content by AES. The slag was analyzed by X-ray diffraction (XRD) to analyze reaction product types. Finally, the microstructure of the alloys at the end of the process was analyzed by scanning electron microscopy (SEM).

3. Results and discussion

Figure 2 shows the relationship of the magnesium content measured by AES with respect to the treatment time from 1 to 60 minutes, at 750, 800, and 850°C and

wt. %	Si	Fe	Cu	Mn	Mg	Cr	Ni	Zn	Ti	Otros	Al
Aluminum scrap	0.286	0.625	0.188	0.77	0.93	0.022	0.013	0.15	0.017	0.024	96.7
ZnO powder	<0.01	0.02	—	0.01	—	—	—	76.65	—	22.32	96.7

Table 1.
Chemical composition of aluminum scrap and ZnO used by aluminothermic reduction process.

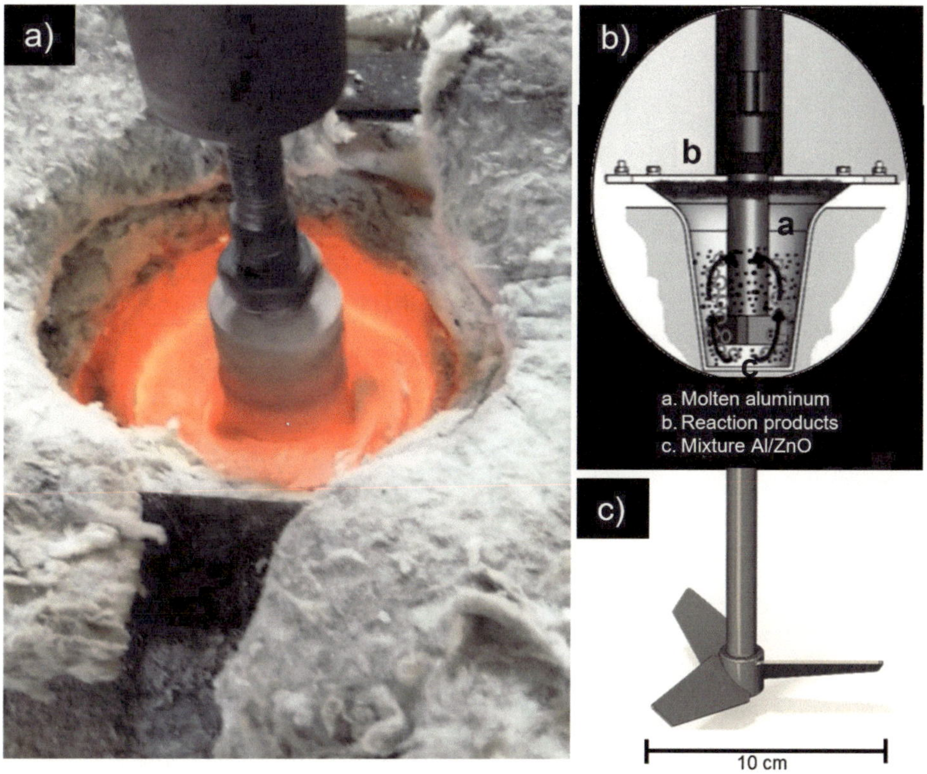

Figure 1.
(a) Top view photography of furnace, (b) scheme of the interior furnace, and (c) the impeller used during the experimentation in aluminothermic reduction process.

stirring velocity of the impeller was 200, 250, and 300 rpm. The results show a significant decrease in magnesium content for all conditions of work. **Figure 3a–c** shows the increase of zinc content in the experiments. Therefore, it can be deduced that the reaction of Eq. (1) governed the process of magnesium removal and zinc added in melted aluminum. The temperature is the variable with the strongest influence on the process, no matter the agitation speed, observing better results at 800 and 850°C, while at 750°C the contents of magnesium and Zn were out the specification for to be used in the automotive industry.

$$1/2\ Al_{(l)} + 1/4\ Mg_{(l)} + ZnO_{(s)} \rightarrow 1/4\ MgAl_2O_{4(s)} + Zn_{(l)}\ \Delta G^0 = -223.99\ kJ \quad (1)$$

Graphics of **Figures 2** and **3** indicate that the processes have an exponential reaction after 20 minutes because this is the time necessary for wettability of the ZnO solid particle with molten aluminum, which allows the reaction to be carried out, where magnesium has an important effect [13, 14]. **Figure 3d** shows the ratio in wt% of magnesium and zinc content at 800°C and 250 rpm, this experiment was more favorable in the present study, because both contents are in the specification for automotive application, giving values <0.06 wt% of Mg and 5.52 wt% of zinc. Other zinc contents obtained were 1.66, 2.3, 2.3, 3.2, and 4 wt%, values used in the aluminum industry in the 700 series alloys. Thus, the reduction process of ZnO powder can remove magnesium and incorporate zinc in molten aluminum.

According to Hashiguchi [15], the change in the magnesium content can also be to the reaction between dissolved magnesium and ambient oxygen gas too, the process is carried out at ambient conditions can be explained by the reaction of

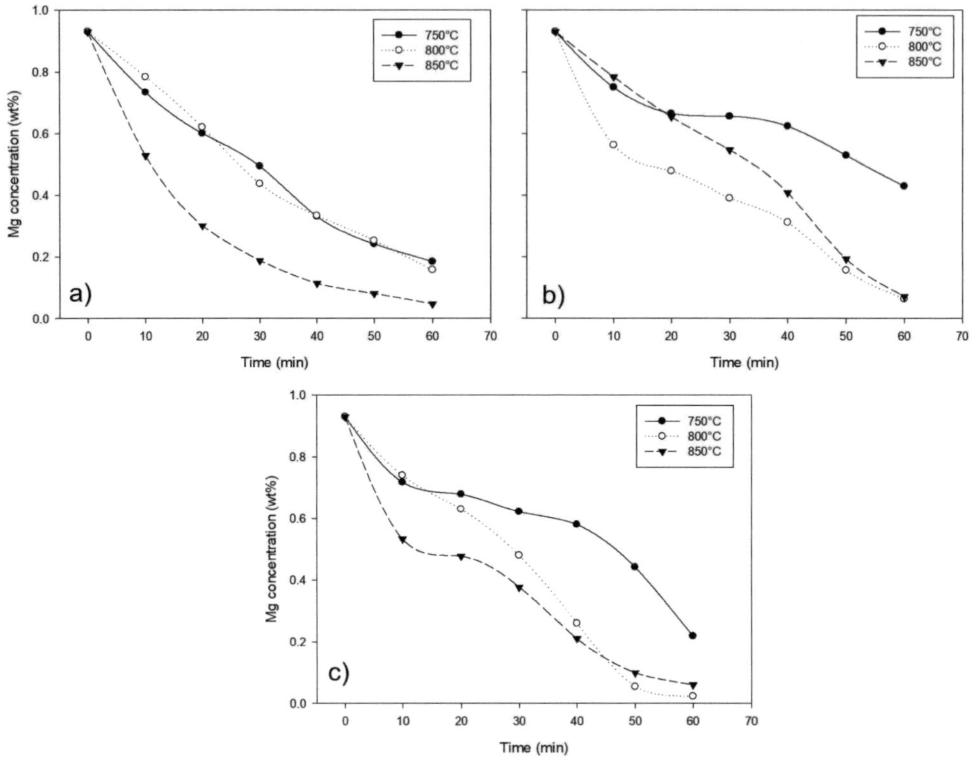

Figure 2.
Relationship between magnesium content and time with respect to temperature and stirring velocity, (a) 200, (b) 250, and (c) 300 rpm.

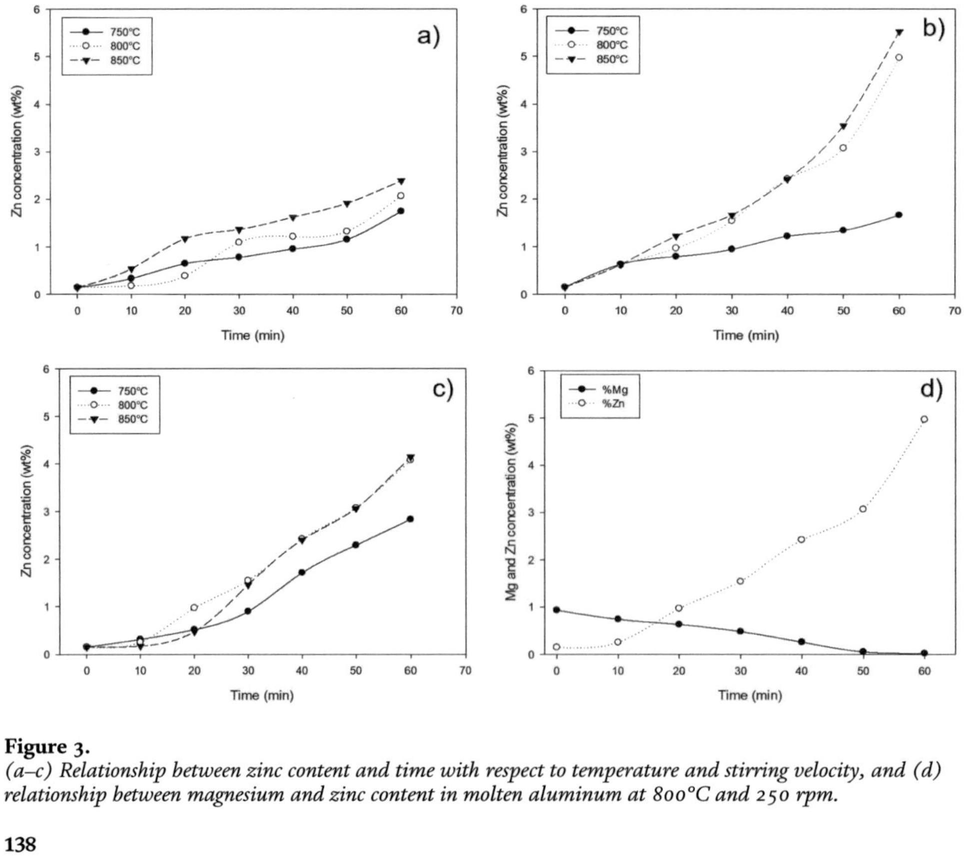

Figure 3.
(a–c) Relationship between zinc content and time with respect to temperature and stirring velocity, and (d) relationship between magnesium and zinc content in molten aluminum at 800°C and 250 rpm.

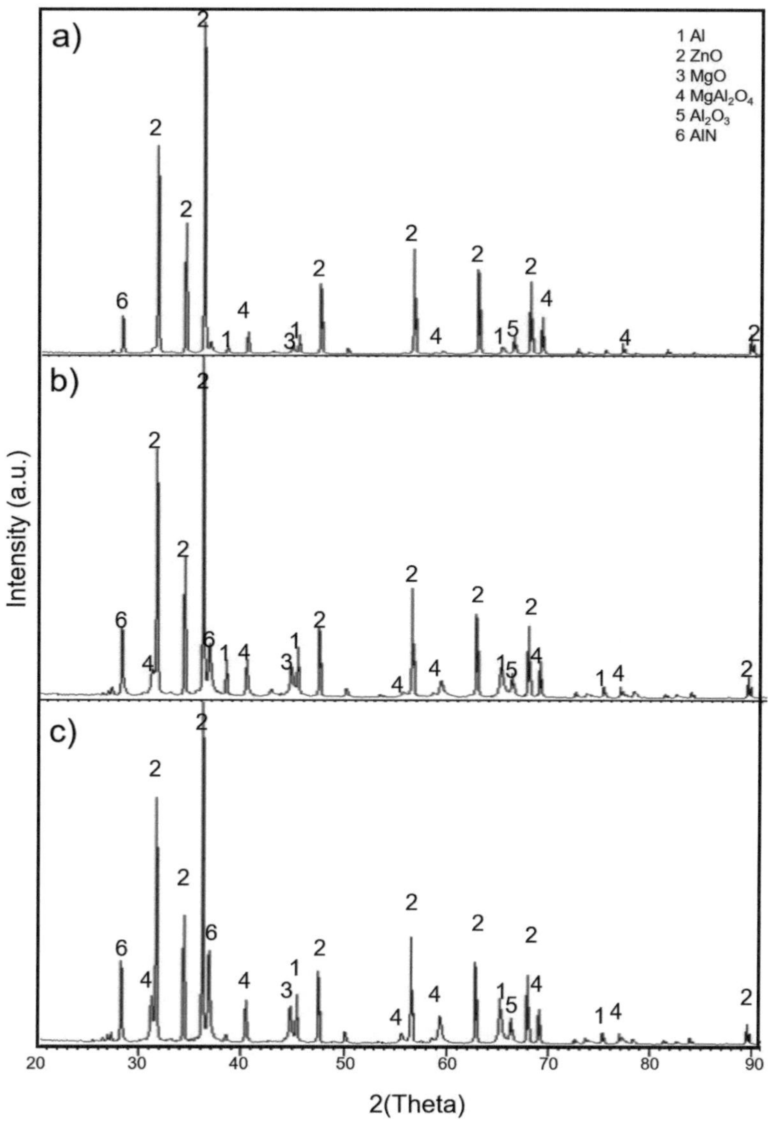

Figure 4.
XRD pattern of the slags produced at 250 rpm and (a) 750°C, (b) 800°C, and (c) 850°C during aluminothermic reduction of ZnO.

Eq. (2). This can be demonstrated by the XRD analysis in **Figure 4**, where the slags obtained in the experiments at 250 rpm and at 750, 800, 850°C are presented, the compounds of these results are $MgAl_2O_4$, MgO, and Al_2O_3. Furthermore, it has been reported that the reactions involved between solid ZnO and molten Al-Mg can be carried out by the Eqs. (1)–(9), values of free energy calculated at 850°C.

$$Mg_{(l)} + O_{2(g)} \rightarrow MgO_{(s)}\ \Delta G^0 = -970.09\ kJ \tag{2}$$

$$4/3\ Al_{2(l)} + O_{2(g)} \rightarrow 2/3Al_2O_{3(s)}\ \Delta G^0 = -882.46\ kJ \tag{3}$$

$$Al_2O_{3(s)} + Mg_{(l)} + 1/2O_{2(g)} = MgAl_2O_{4(s)}\ \Delta G^0 = -1036.56\ kJ \tag{4}$$

$$2/3\ Al_{(l)} + 1/3\ MgO_{(s)} + ZnO_{(s)} \rightarrow 1/3\ MgAl_2O_{4(s)} + Zn_{(l)}\ \Delta G^0 = -220.23\ kJ \tag{5}$$

$$4/3\ Al_{(l)} + 2ZnO_{(s)} = 2/3\ Al_2O_{3(s)} + 2Zn_{(l)}\ \Delta G^0 = -205.82\ kJ \tag{6}$$

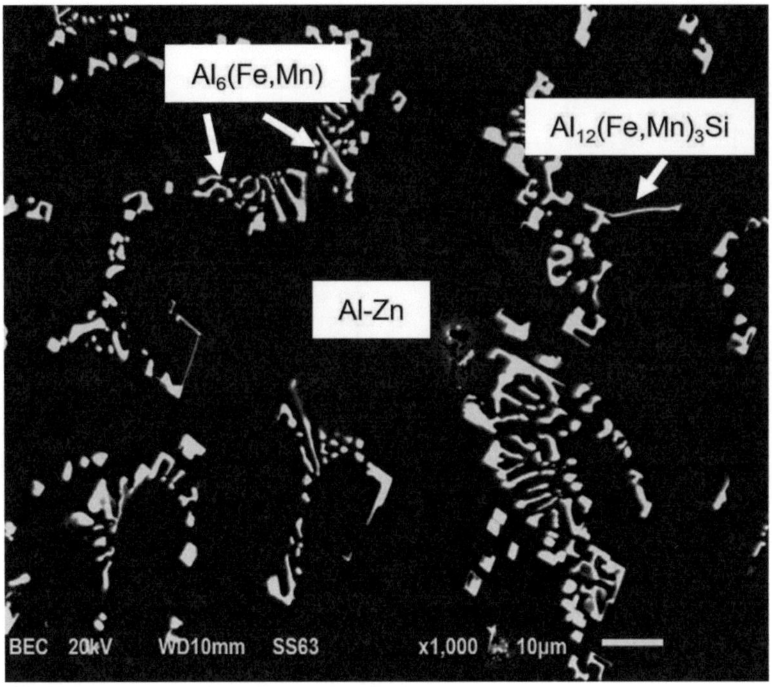

Figure 5.
SEM images of Al-Zn alloys from aluminothermic reduction process at 250 rpm and (a) 750°C, (b) 800°C, and (c) 850°C. (d) Al_6(Fe,Mn) and $Al_{12}(Fe,Mn)_3Si$ phases on Al-Zn matrix.

$$Mg_{(l)} + ZnO_{(s)} \rightarrow MgO_{(s)} + Zn_{(l)}\ \Delta G^0 = -244.91\ kJ \quad (7)$$

$$Al_2O_{3(s)} + 3Mg_{(l)} \rightarrow 3MgO_{(s)} + 2Al_{(l)}\ \Delta G^0 = -117.69\ kJ \quad (8)$$

$$Al_2O_{3(s)} + MgO_{(s)} \rightarrow MgAl_2O_{4(s)}\ \Delta G^0 = -92.02\ kJ \quad (9)$$

According to the free energy values of each reaction calculated at 850°C, the decrease of magnesium and the increase of Zn in the aluminothermic reduction process can be carried out mainly by reactions of the Eqs. (1), (2), and (4)–(7) while Eqs. (3), (8), and (9) are reactions for slag formation.

The beverage cans used in this work were without any type of refining treatment, therefore, the alloy presented higher contents of Fe, Si, Mn. **Figure 5** shows microstructure after the aluminothermic reduction process of secondary aluminum with a matrix rich in zinc and second phases of Al_6(Fe, Mn) and $Al_{12}(Fe, Mn)_3Si$, the first phase with massive blocky morphology due to the content of the Mn in the alloy, according to Liu [16], this element is the one that determines the morphology of this type of intermetallic, while the phase $Al_{12}(Fe, Mn)_3Si$ was presented less quantity, this phase is formed by the cooling rate and diffusion of Si to the Al_6(Fe, Mn). Previous studies [17, 18] indicate that this transformation can also be carried out in processes where there is a loss of magnesium and Si was effective nucleating substrates for the transformation of Al_6(Fe, Mn) to $Al_{12}(Fe, Mn)_3Si$.

4. Conclusion

Magnesium was removed from the aluminum scrap using the aluminothermic reduction process of ZnO powder. The temperatures favored the decrease of magnesium in the metal bath and generated reaction products as $MgAl_2O_4$, MgO, and Al_2O_3 in solid-state, so that this process can help to reduce the emission of polluting gas. Also, the ZnO powder was reduced, and the zinc incorporated into the melted aluminum microstructure of the alloys has an Al-Zn matrix and second phases of Al_6(Fe, Mn) and Al_{12}(Fe, Mn)$_3$Si in the cast state. In this case, this process can be favorable for the manufacture of alloys that can be applied in the automotive sector.

Acknowledgements

The authors thank CONACYT for financial support and institutions as CINVESTAV-Saltillo and COMIMSA-Saltillo for allowing the use of their laboratories.

Author details

Rocio Maricela Ochoa Palacios[1*], Citlaly Castillo Rodriguez[1], Jesus Torres Torres[2], Perla Janet Resendiz Hernandez[1] and Alfredo Flores Valdes[2]

1 Tecnologico Nacional de México/IT de Saltillo, Saltillo, Coahuila, México

2 Center for Research and Advanced Studies of the National Polytechnic Institute Metallurgical Engineering, Gustavo A. Madero, Mexico

*Address all correspondence to: rochoa@itsaltillo.edu.mx

IntechOpen

© 2022 The Author(s). Licensee IntechOpen. This chapter is distributed under the terms of the Creative Commons Attribution License (http://creativecommons.org/licenses/by/3.0), which permits unrestricted use, distribution, and reproduction in any medium, provided the original work is properly cited.

References

[1] Bojun X. Optimization of Recycling Process of Die Cast Aluminum A380 Machining Chips [thesis]. Ontario, Canada: University of Windsor; 2015

[2] Van der Harst E, Potting J, Kroeze C. Comparison of different methods to include recycling in LCAs of aluminium cans and disposable polystyrene cups. Waste Management. 2018;**48**:565-583. DOI: 10.1016/j.wasman.2015.09.027

[3] Stotz PM, Niero M, Bey N, Paraskevas D. Environmental screening of novel technologies to increase material circularity: A case study on aluminium cans. Resources Conservation Recycling. 2017;**127**: 96-106. DOI: 10.1016/j.resconrec.2017.07.013

[4] Capuzzi S, Timelli G. Preparation and melting of scrap in aluminum recycling: A review. Metals. 2018; **8**(249):2-24. DOI: 10.3390/met8040249

[5] David E, Kopac J. Use of separation and impurity removal methods to improve aluminium waste recycling process. Materials Today: Proceedings. 2015;**2**(10):5071-5079. DOI: 10.1016/j.matpr.2015.10.098

[6] Gray AR. United States Patent: 3871965; 2010

[7] Briolle E. United States Patent: 5232300; 1993

[8] Hiraki T, Miki T, Nakajima K, Matsubae K, Nakamura S, Nagasaka T. Thermodynamic analysis for the refining ability of salt flux for aluminum recycling. Materials. 2014;**7**(8): 5543-5553. DOI: 10.3390/ma7085543

[9] Kamat GR, Gupta CK. Open aluminothermic reduction of columbium (Nb) pentoxide and purification of the reduced metal. Metallurgical and Transactions. 1971; **2**(10):2817-2823. DOI: 10.1007/bf02813257

[10] Lin N, Han Y, Zhou J, Zhang K, Xu T, Qiuan Y. A low temperature molten salt process for aluminothermic reduction of silicon oxides to crystalline Si for Li-ion batteries. Energy and Environmental Science. 2015;**8**(11): 3187-3191. DOI: 10.1039/C5EE02487K

[11] Zhu HX, Abbaschian R. In-situ processing of NiAl-alumina composites by thermite reaction. Materials Science and Engineering: A. 2000;**282**(1):1-7. DOI: 10.1016/S0921-5093(99)00788-1

[12] Lima BB, Ramos AS, Nunes CA, Conte RA. Ni-65wt%Nb alloy by aluminothermic reduction process. International Journal of Refractory Metals and Hard Materials. 2000;**8**(6): 267-271. DOI: 10.1016/S0263-4368(00)00021-4

[13] Zhang L, Lv X, Torgerson AT, Long M. Removal of impurity elements from molten aluminum: A review. Mineral Processing and Extractive Metallurgy Review. 2011;**32**(3):150-228. DOI: 10.1080/08827508.2010.483396

[14] Muñiz R, Flores A, Torres J, Luna S, Rodríguez N. A kinetic study of the strontium extraction by metallothermic reduction using submerged SrO powders injection. Materials Letters. 2008;**62**(4–5):637-640. DOI: 10.1016/j.matlet.2007.06.024

[15] Hashiguchi T, Sueyoshi H. Removal of magnesium from molten aluminum scrap by compound-separation method with Shirasu. Materials Transactions. 2010;**51**(5):838-843. DOI: 10.2320/matertrans.mh200914

[16] Liu Y, Luo L, Han C, Ou L, Wang J. Effect of Fe, Si and cooling rate on the formation of Fe- and Mn-rich intermetallics in Al–5Mg–0.8Mn Alloy.

Journal of Materials Science & Technology. 2016;**32**(4):305-312. DOI: 10.1016/j.jmst.2015.10.010

[17] Kamat R, Ng-Yelim J, Saimoto S. Morphology and precipitation of α-Al (Fe, Mn)Si phase in hot rolled AA3004. Journal of Materials Research. 1995; **86**(1):49-53. DOI: 10.1515/ijmr-1995-860110

[18] Anyalebechi PN, Rouns T, Sanders RE. Microstructural analysis of second phase developed during casting and preheating of 3XXX aluminum alloys. In: Hirano K, Oikawa H, Ikeda K, editors. Proceedings of RASELM 91, Conference on Science and Engineering of Light Metals. Tokyo, Japan: Institute of Light Metals; 1991. p. 923

Chapter 9

Mechanical Resistance of a Superficially Treated Alloy Drill Pipe during Onshore Drilling

Lallia Belkacem

Abstract

A theoretical study was conducted to investigate the limit depth reached by aluminum alloy drill pipes combined with steel pipes for deep well drilling in the Algerian country. Therefore, the present study is based on various parameters that have an impact on the fatigue behavior of these tubes, focusing particularly on the damage caused by the gravity of the dogleg in the crooked path in addition to their mechanical behavior, and determines the extent to which the aluminum drill pipe can drill without failures using well-engineering software modeling which provides the expected loads in the drilling and provides results more or less close to reality. This analysis indicates that the aluminum drill pipe has good fatigue resistance, despite the cumulative presence of axial and bending stress concentrations in the dogleg zones.

Keywords: tortuous trajectory, torque, drag, dogleg, 2024 aluminum alloy, fatigue

1. Introduction

The increased depth of drilling over 4500 m has affected the complexity of drilling operation as well as drilling rigs and their practices. Due to this, modern drilling technologies are designed to increase the capacity of drilling equipment. Because the possibility of increasing the drive power is quite limited and economically impractical, the application of alternative materials such as aluminum alloys becomes the most cost-effective and relevant target [1]. For instance, drilling deep, ultra-deep, and especially horizontal wells is extremely important to ensure the high operational reliability of drill string, to reduce its stress-strain state, and to ensure trouble-free operation at extreme loads and high temperatures. Drill string assembly and its weight significantly affect the technical and economic parameters of the well drilling process, the behavior of resisting forces, and specify the level of load in the parts of a drilling rig.

Aluminum alloys possess several valuable physical and mechanical properties that favorably distinguish them from steel, which is a traditional material in the drill pipe. The following should be referred to as the basic properties of aluminum alloys:

- Low specific weight;
- Reduced modulus of longitudinal elasticity and shear;

- Workability in pipe production using extrusion process;
- Corrosion resistance in aggressive environment and, first of all, in H_2S and CO_2.

Aluminum alloys are still of relatively limited use in drilling, since the temperature is of crucial importance here, since in deep wells, depending on the geological section, the temperature at 3500–7000 m can reach 42,315 K, and in some cases 52,315–82,315 K [2, 3] for example D16 and 1953 alloys. D16 alloy has high strength characteristics at room and elevated (up to 43,315 K) temperatures, medium ductility, but very low corrosion resistance in drilling fluids.

- Non-magnetic and vibration resistance properties;
- High rate of weight reduction in solutions of various density [1, 4].

These properties of aluminum alloys specify the basic efficiency in light alloy aluminum drill pipe (LAIDP) application in drill strings while constructing oil and gas wells. Therefore, the work process in this paper is to first determine the limit up to which the aluminum drill pipe can drill without failures by means of the Decision Space Well Engineering software and second, point out the expected loads during drilling [5].

2. Superficially treated 2024 aluminum alloy technology

The method of oxidization by micro arcs is an electrochemical surface treatment; it is similar to the anodization, where we use a very high voltage in order to produce electric discharges. It is a chemical conversion of the substrate by its oxide, therefore an excellent adhesion. The oxide layer is formed by applying an electric potential (at least 200 V), whereas the piece is immersed in an acid electrolyte. Discharges are produced, agglomerate, and make the oxide dense and partially converted to amorphous alumina in crystal form such as the cordon. The structure of coatings applied by the micro-arc oxidation method to DI6 aluminum alloy) was investigated with the use of scanning electron microscopy and chemical and X-ray diffraction analyses. It was found that the coating is inhomogeneous. Its structure and composition change across the thickness and the micro inhomogeneity observed in analyzing the microstructure of a transverse micro-sample and confirmed by the data of coating microhardness measurements are characterized by the formation of a honeycomb type structure. Consequently, the coating may be considered as a composition material with uniformly distributed hard islands surrounded by softer veins. **Tables 1** and **2** shown chemical composition and mechanical characteristics of aluminum alloy 2024 [5].

Designation (NF A03104)	Chemical composition (%)					
2024	AL	Cu	Mn	Mg	Si	Fe
	The complement	3.8–4.5	0.3–4.5	1.2–1.8	0.5	0.5

Table 1.
Chemical composition of the superficially treated alloy.

Tube sections	Ultimate tensile strength σr [MPa]	Stress σ0,2 in [MPa]	Elasticity limit σpr [MPa]	Relative extension to rupture δ en %	elastic range slope α [°]	Young's modulus E [MPa]
2 × 30	Before superficial treatment (OMA)					
	491	382	360	12	80	71,805
	After superficial treatment (OMA)					
	468	360	337	13	81	8763

Table 2.
Results of tensile testing (micro-tests) on test-tubes in the superficially treated alloy, carried out on the machine "ALA-TOO".

3. Stress loads

The Decision Space Well Engineering Software Torque & Drag outputs can be used to predict and analyze the torque and axial forces generated by drill strings, casing strings, or liners while running in, pulling out, sliding, backreaming, and/or rotating in a three-dimensional wellbore. The effects of mud properties, wellbore deviation, weight-on-bit (WOB), and other operational parameters can be studied.

In Decision Space Well Engineering, many stress calculations are performed using the following equations. These calculations include the effect of [6]:

- Axial stress (tension + bending + buckling)
- Bending stress approximated from wellbore curvature
- Bending stress due to buckling
- Hoop stress due to internal and external pressure
- Radial stress due to internal and external pressure
- Torsional stress from twist
- Von Misses force.

3.1 Drilling for 8½"hole section with steel drill pipe

The root causes for different problems encountered on the 8½ section in well drilled in Algeria are;

1. Hole geometry (wellbore tortuosity 0.2032 m).
2. Inharmonic in drilling string and parameters: weight-on-bit (WOB), rotational speed (RPM), bottom hole assembly (BHA).
3. High torque peaks, while drilling and back reaming (37,962–48,809) Nm recorded at the surface [3].

3.2 Load summary

According to **Table 3**, the drill pipe is exposed to fatigue failure during all operations, which is naturally related to the well trajectory, which has some

Load condition	Stress failure	Measured weight (tonne)	Total stretch (m)	Axial stress = 0	
				Distance from surface (m)	Distance from bit (m)
Tripping in	Fatigue	123	6.60	3793	208
Tripping out		161	7.63	3793	208
Rotating on bottom		137	6.74	3793	208
Backreaming		149	7.36	3793	208
Rotating off bottom		147	7.25	3793	208

- Tripping in: when running in hole (RIH)
- Tripping out: when pull out hole (POOH)
- Rotating on bottom: means that the pipe rotates with axial movement.
- Rotating off bottom: means that the pipe rotates without any axial movement
- Backreaming: the practice of pumping and rotating the drill string while simultaneously pulling out of the hole.

Table 3.
Load summary for steel drill pipe.

tortuosity as shown in **Table 4**, which displays doglegs per unit length in the string during drilling operations and located points along the well that may be subjected to high severe doglegs leading to a high degree of severity of the moment bending on the mentioned areas. For that reason; the solution proposed in this paper consists in simply replacing joints of standard steel drill pipe with lighter aluminum drill pipe while keeping the same bottom hole assembly (BHA).

Thus; A torque and drag optimization study has been run to determine the optimum number of ADP joints along the drill string to minimize friction and reduce compression along the well trajectory and limit the maximum depth which can aluminum drill pipe reached without any failure, accordingly, 150 joints of standard steel DP have been replaced by aluminum DP above the BHA. **Table 5** represented below give the main characteristics of ADP.

3.3 Results and interpretation

The results exposed in **Table 6** indicate that aluminum drill pipe gives good results in tortuous intervals compared to steel drill pipe but in total depth do not exceed the 4000 m.

This result is confirmed by the output obtained in the torque drag effective tension graph which has included the graphical curves on tension vs. distance along with string (tension limit, helical buckling, sinusoidal buckling, in all operations (rotate off bottom, rotate on bottom, tripping out, tripping in).

Accordingly; it is obviously seen that all operations curves do not cross the Tension Limit curve; the drill string is located into a safe window; therefore, there is no danger of exertion the aluminum drill pipe in the tortuous interval, as depicted in **Figure 1**.

Thus we can assume that the 0.2032 m drilling phase benefited from ADP for the reason that; First of all aluminum drill pipe has significantly good fatigue, which is confirmed by the fatigue ratio value which is 0.642 less than 1. As shown on **Figure 2**.

$$\text{Fatigue ratio } R_f = \frac{\left(\left|\sigma_{\text{bending}}\right| + \left|\sigma_{\text{buckling}}\right|\right)}{\sigma_{\text{fatigue Limt}}} = 1 \text{ is equal to the safety limit} \quad (1)$$

where $\sigma_{\text{bendiing}} = \sigma_{\text{flexion}}$ [6].

DOI: http://dx.doi.org/10.5772/intechopen.101867

MD (m) measured depth	Inclination (°)	AZ (°)	TVD (m) Vertical depth	DLS (°/30 m) Dog leg	Absolut tortuosity (°/30 m)	Reel tortuosity / (°30 m)	Vertical section (m)	Build (°/30 m)	Walk (°/30 m)
2905.00	1.00	63.67	2905.22	0.010	0.010	0.000	11.25	0.010	0.000
2912.00	1.14	57.89	2912.37	0.741	0.012	0.000	11.31	0.587	−24.252
2921.00	1.03	66.98	2921.51	0.669	0.014	0.000	11.39	−0.361	29.836
2930.80	1.05	116.65	2930.65	2.868	0.023	0.000	11.39	0.066	163.031
2939.00	1.63	161.92	2939.70	3.852	0.035	0.000	11.23	1.923	150.066
2949.00	3.42	180.29	2948.93	6.306	0.055	0.000	10.83	5.812	59.643
2958.00	5.52	183.53	2958.05	6.932	0.076	0.000	10.11	6.885	10.623
2967.38	6.72	183.90	2967.14	3.941	0.088	0.000	9.14	3.939	1.214
2976.52	7.50	185.52	2976.21	2.643	0.096	0.000	8.01	2.560	5.317
2985.67	7.92	186.26	2985.27	1.415	0.100	0.000	6.79	1.377	2.426
2994.81	8.17	186.80	2994.32	0.857	0.102	0.000	5.52	0.821	1.772
3003.96	8.44	187.19	3003.38	0.904	0.104	0.000	4.21	0.885	1.279
3013.10	8.44	189.80	3012.42	1.257	0.108	0.000	2.88	0.000	8.567
3022.24	8.55	191.52	3021.46	0.909	0.110	0.000	1.56	0.361	5.646
3031.39	8.31	193.54	3030.51	1.250	0.114	0.000	0.25	−0.787	6.623
3040.53	8.33	194.91	3039.55	0.654	0.115	0.000	−1.03	0.066	4.497
3049.68	8.34	196.13	3048.61	0.581	0.117	0.000	−2.31	0.033	4.000
3058.82	7.82	193.58	3057.66	2.073	0.123	0.000	−3.55	−1.707	−8.370
3067.96	7.22	203.80	3066.72	4.803	0.137	0.000	−4.68	−1.969	33.545
3077.11	9.00	206.56	3075.78	5.972	0.154	0.000	−5.85	5.836	9.049
3086.25	5.96	209.55	3084.84	10.056	0.183	0.000	−6.90	−9.978	9.814

MD (m) measured depth	Inclination (°)	AZ (°)	TVD (m) Vertical depth	DLS (°/30 m) Dog leg	Absolut tortuosity (°/30 m)	Reel tortuosity / (°30 m)	Vertical section (m)	Build (°/30 m)	Walk (°/30 m)
3095.40	5.24	213.98	3093.94	2.752	0.191	0.000	−7.66	−2.361	14.525
3113.68	4.23	223.24	3112.16	2.074	0.202	0.000	−8.85	−1.658	15.197
3122.83	4.09	225.14	3121.29	0.644	0.203	0.000	−9.32	−0.459	6.230
3131.97	4.90	228.33	3130.40	2.781	0.211	0.000	−9.81	2.659	10.470
3177.69	1.70	243.22	3176.04	2.156	0.239	0.000	−11.42	−2.100	9.770
3195.98	0.87	265.04	3194.32	1.557	0.246	0.000	−11.55	−1.361	35.790
4000.65	0.94	232.89	3998.90	0.019	0.200	0.000	−16.06	0.003	−1.199

Table 4.
Variation in wall trajectory via tortuosity.

Hole section 0.2032 m	Steel drill pipe/aluminum	
	Drill pipe (0.127 m)-G105	**DP drill pipe (0.127 m) -AL2024**
Depth	1350 (m)	2592.86 (m)
Elastic limit σe (MPa)	724	52127.5
Tensile strength σ m (MPa))	931	359
Young's modulus of elasticity E (MPa)	206896.55	88,763
σ endurance limite (MPa)	137.89	160

Note: The tool-joints for aluminum drill pipe are manufactured from steel.

Table 5.
Details in each specific drilling string and section.

Load condition	Stress failure	Measured weight (tonne)	Total stretch (m)	Axial stress = 0	
	Fatigue			**Distance from surface (m)**	**Distance from bit (m)**
Tripping in		121,560	7.31	3793	207
Tripping out		160,340	8.52	3793	207
Rotating on bottom		135,690	7.34	3793	207
Backreaming		147,680	8.21	3793	207
Rotating off bottom		14,680	8.07	3793	207

- *Tripping in: when running in hole (RIH)*
-*Tripping out: when pull out hole (POOH)*
- *Rotating on bottom: means that the pipe rotates with axial movement.*
-*Rotating off bottom: means that the pipe rotates without any axial movement*
- *Backreaming: the practice of pumping and rotating the drill string while simultaneously pulling out of the hole.*

Table 6.
Load summary for aluminum drill pipe.

Secondly; the 0.2032 m drilling phase benefited from (aluminum drill pipe) ADP since there was a considerable reduction of the surface torque and hook load as illustrated in **Figure 3**, for the reason that there was a reduction in side force as represented in **Table** 7.

As a final point; analysis of stresses in drill pipe and their results on each one, are shown on the graph cited, in **Figure 4** which gives us the different results for the Stresses (psi), the critically of Failure will depend on the value of these stresses. The Von Mises yield condition, states initial yield limit is based on the combination of the three principal stresses axial stress, radial stress, and hoop stress [3].

$$\sigma_{VM} = \sqrt{\frac{(\sigma_r - \sigma_h{}^2) + (\sigma_a - \sigma_r)^2 + (\sigma_h - \sigma_a)^2 + 6\sigma_s^2 + 6\sigma_t^2}{2}} \tag{2}$$

σ_{VM}: Von misses stresses, σ_a: axial stress, σ_r: radial stress, σ_h: hoop stress, σ_t: torsion stress σ_S: shear stress [6].

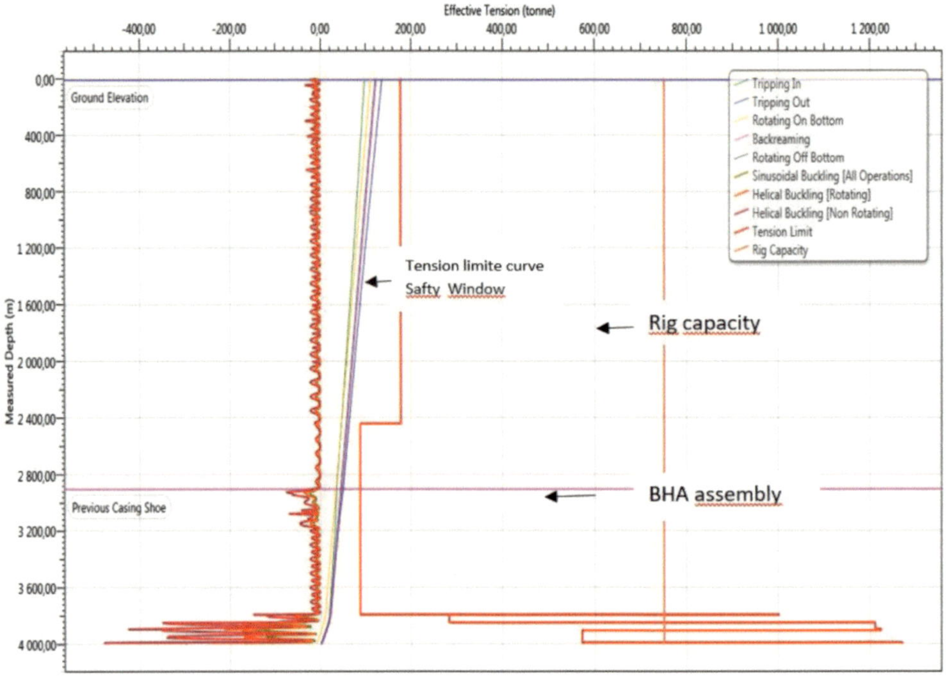

Figure 1.
Effective tension limit curve.

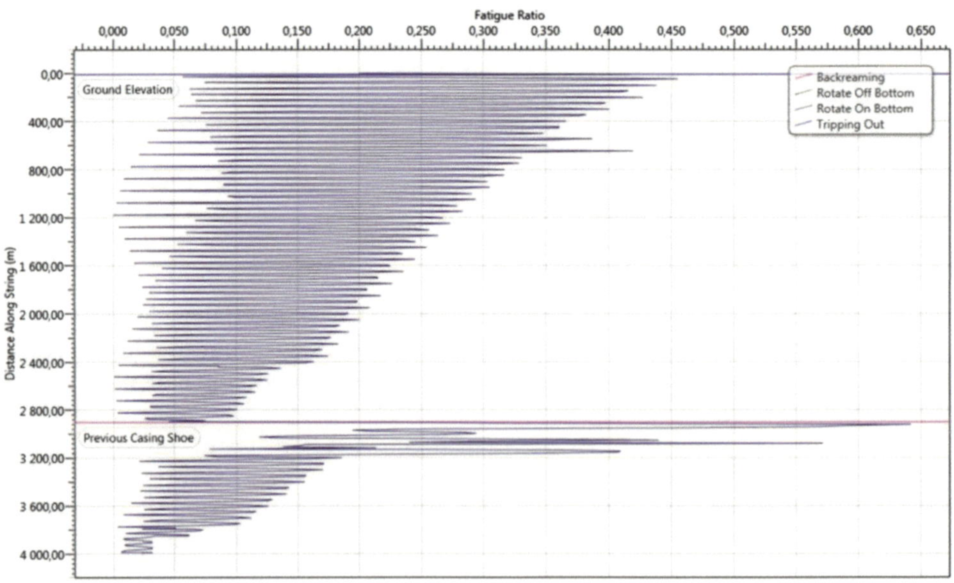

Figure 2.
Aluminum drill pipe fatigue ration curve.

Thus; we can be well observed that the Von-Misses Stress is under the tension limit line (red color), due to our model simulation is located into the safety window.

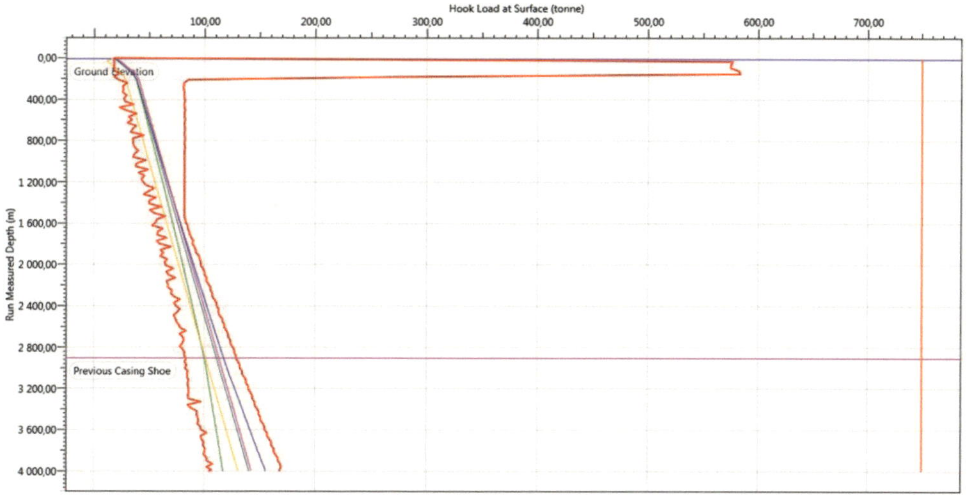

Figure 3.
Hook load for aluminum drill pipe (ADP).

	Rotating on bottom ADP side force	Rotating on bottom SDP side force
(m)	**(N)**	**(lbf/length)**
2908	3114	3172
2917	5805	5907
2926	10,814	11,009
2935	958	9751
2944	8723	8874
2953	6135	6326
2962	3177	3230
2971	2532	2572
2981	2736	2785
2990	3759	3822
2999	3906	3968
3008	3501	3559
3017	2024	2059
3026	1940	1975
3035	2296	2335
3044	4622	4706
3062	4422	4493
3071	3684	3750
30,791	9662	9831

Table 7.
Side force aluminum drill pipe (ADP)/steel drill pipe (DP).

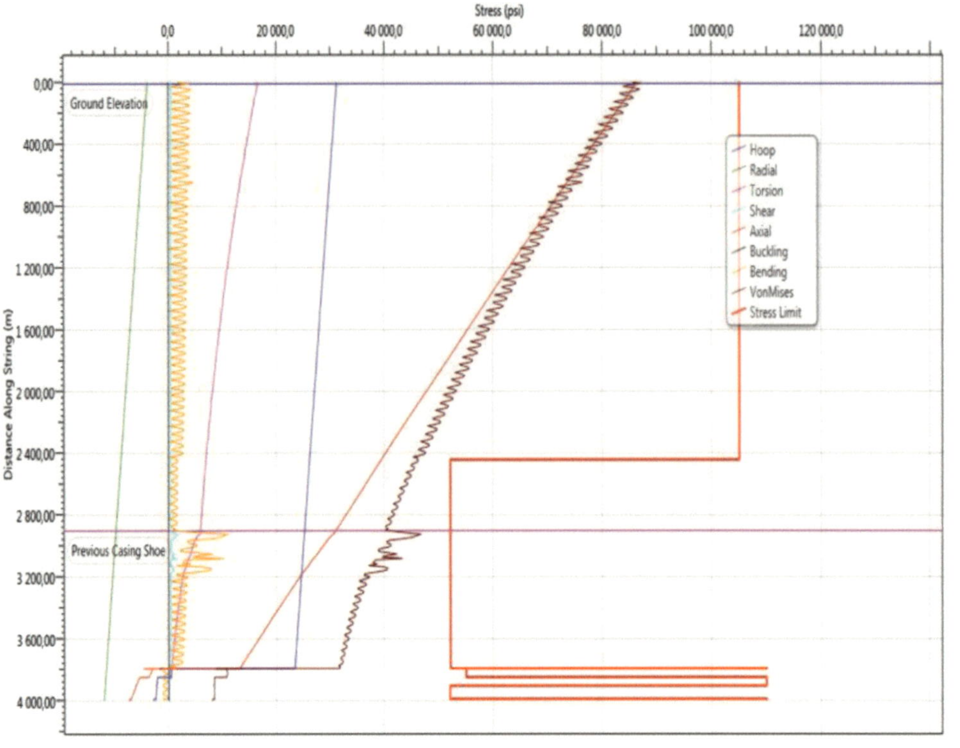

Figure 4.
Aluminum drill pipe tripping out stress detail.

4. Conclusion

In synopsis; aluminum drill pipe is subjected to less strain than steel drill pipe. This is due to their lower modulus of elasticity and lighter density than steel drill pipe. Which leads to a good fatigue resistance even at the simultaneous presence of high applied torque and axial load, and severe doglegs, which makes the use of other options Steel drill pipe impractical from the fatigue failure standpoint. The limiting factor for using aluminum drill pipe in such severe conditions is not the fatigue resistance of the material that the pipe is made of, but the high contact forces between the wellbore and aluminum drill pipe.

Author details

Lallia Belkacem
Faculty of Hydrocarbons and Chemistry Independence Street, Laboratory of Petroleum Equipments Reliability and Materials, University Mohamed Bougara, Boumerdes, Algeria

*Address all correspondence to: l.belkacem@univ-boumerdes.dz

IntechOpen

© 2022 The Author(s). Licensee IntechOpen. This chapter is distributed under the terms of the Creative Commons Attribution License (http://creativecommons.org/licenses/by/3.0), which permits unrestricted use, distribution, and reproduction in any medium, provided the original work is properly cited.

References

[1] Menand S, LehnerJ JK. Successful use of mixed aluminum-steel drill pipe string in complex horizontal wells. Journal of Petroleum Science and Engineering. 2014;**SPE-170255**:13

[2] Aleksandrov VS, Abdel'baki N, Fedorov VA, Kan AG. Investigation of the crack formation process in the hardened layer obtained on D16 alloy by the microarc oxidation method in static loading Chemical and Petroleum Engineering. 1990;**26**:586–587

[3] Daily drilling report of 8 ½" hole section of well in Algeria field. 2015

[4] Gelfgat MY, Basovich VS, Tikhonov VS. Drill string with aluminum alloy pipes design and practices. Paper SPE 79873. 2003

[5] Aleksandrov V, Kan A, Abdel'baki N, et al. The influence of surface treatment of DI6 alloy by the method of microarc oxidation on its strength properties in static loading. Chemical and Petroleum Engineering. 1988;**24**(9):493-495

[6] Decision Space Well Engineering Software. EDT_5000.14.00 Software system. Landmark products

Chapter 10

How Impact the Design of Aluminum Swaging Circle Fitting on the Sealing for Piping Systems: Analytical and Numerical Model

Ahmet Atak

Abstract

Installation and repair of hydraulic installations are carried out by joining the pipes in the field. Pipe connections in aviation are made in a very narrow space and field. For this reason, fitting swaging method is used to connect the fittings to the pipes with a hydraulic hand tool. The basis for developing a swaging tool is knowledge of the design parameters for the plastic deformation of the swaging circle fitting. In addition to the design parameter, pipes have to be joined in cryogenic vacuum conditions, which require sealing in such sensitive and harsh conditions. In this study, the effect of swaged circle fitting designs on tightness and strength was determined and different swaging methods were examined for its improvement. Different geometric swaged circle fittings are designed and analytical and numerical models are solved. The solution results obtained show the characteristic effect of the fitting swaging analysis methodology and the swaged circle fitting design on the sealing of the pipe joint. The innovation is mainly the effect of the swaging circle connection design on the sealing of the pipe joint. With the finding in this work, it is now possible to develop or develop new tools for engineered swaging circle fitting.

Keywords: pipe joining, sleeve design, swaging ring, sleeve swaging method, pipe connection

1. Introduction

The structural engineers faced many challenges to safely and durably design and build the budget-consuming projects. The loadings may damage the piping structures or disturb their normal operations whenever their magnitudes reach the strength limits of the structure material [1].

Mostly the structural engineers performed three-dimensional (3D) finite-element analyses to investigate the behavior of buried pipe subject to strike-slip fault movement in dry sand and, more realistically, in partially saturated sand [2].

Some literature presents specific methods and algorithms for evaluating potential damage zones of pipe joints that can be used to make decisions to ensure the safety of use of hazardous production systems that also concern human life in the gas-oil sector during the draft and detailed design and later use stages [3].

IntechOpen

Design analysis engineers recommend to specify mechanical closure and fatigue conditions considering stress and amplitude depending on swage parameter and material for using swaged fittings to joining in structural engineering and mechanical handling [4].

In some load cases, to obtain more realistic results from design analysis for structural systems, it is necessary to determine or to define suitable fastener stiffness values in their connections [5].

It is necessary to propose an analytical or numerical useful method for the practicing engineers in the rational design of pipe connections in detail designing and structural analysis [6].

The most critically part of a piping is they are connections, each other or to equipment. Piping solutions using non-welded connections and cold bent piping offer significant value through reductions in fabrication and commissioning time, while improving workplace safety. The benefits of non-welded piping technology are ranging from reduced preparation and inspection time to a safer work environment [7].

Different methods are used to join the tubes. One of them is rotary swaging, in them the rotation energy changed to thermal friction energy and the fitting is swaged on the pipe with approximately 100% sealing [8].

The swaging as a joining method is used today in most crucial industries including the military, automotive, and medical. Within the military industry, swaging is used to form items such as gun barrels and anti-tank rocket tips. Fittings are also swaged into cable. Countless auto parts and systems, such as distress alert brake cables, steering components, and powertrains, are produced by rolling. In the medical sector; subcutaneous needles, catheter tape assemblies and optical parts and assemblies are the first to come to mind as products produced by rolling the same. In the energy sector; heater elements, heat conductive materials and zirconium profiles can be listed (items that the renewable energy industry relies upon), can be expertly machined using swaging machines. In aerospace; by swaging, the aerospace industry is assured of high-quality control rods, wire rope cable assemblies, and fluid transfer tubing [9].

It is useful for swaging tube and pipe made from the manufacturing industry's most common material (steel) and it's also suitable for stainless steels, aluminum, titanium alloys.

Also, swaged pipe in pipe construction has been increasingly used for offshore pipeline system. The end connection, produced by plastering and rolling, is transmitted to the outer tube by a cold deformation process and then connected to the inner tube. The twin welded piping system provides excellent thermal insulation [7].

In any case, literature studies have shown how important fasteners are to structural design engineers. Piping designs are generally calculated with 16 bar and below, and such connections are designed as removable flange or screw connections. In screw connections in such systems, rubber or metal plastered intermediate element between the screw and the nut provides both the connection and the sealing [10].

Connections in pipe systems operating under high pressure are produced using rubber or metal unions produced using the swaging method. Such connections are designed and tested at 1.5 times the working pressure and at burst pressure. Piping systems used in aerospace are produced from either stainless steel or mostly aluminum alloy pipes. In addition to rubber plastic piping, and also their tightness is tested by performing tests such as vibration tests under conditions far above operating conditions [11].

Although there is not much in this field (ring swaging pipe connection), design and analysis studies have been carried out [12].

As can be seen in the swaging type A in **Figure 1**, both the direct connection of the two pipes and the connection of the fittings with the pipe can be done with the design of the swaging circle fitting (Sleeve), which is one of the main inspection elements of the study. The pipes to be connected can also be of different materials; however, different swaging parameters have to be determined in this case. These connections, as in welded connections, are expected to be made to withstand more pressures than the pipe itself. The design of the sealed swaging circle fitting pipe significantly affects the mechanical strength and tightness of the joint. As can be seen in **Figure 1**, the area to be compacted is designed as inclined and undulated in the axial direction; this prevents it from being pulled in the axial direction. Likewise, the swaging circle fitting creates an obstacle to the tangential rotation of the pipe in the fitting with waves in the circumferential direction. In addition, two internal circumferential grooves have been designed to provide greater sealing by placing an O-Ring in the ring.

In **Figure 1**, a cut-out view of the A-type fitting is given. Compression is applied in two steps by pressing the first pipe from the right and then the pipe from the left towards the center in a radial direction, so that the plastering circle counts the plastic displacements. Pipes and the middle part of the fitting exposed to pressure load does not exceed elastic loading. When pipes are of different materials, different swaging parameters must be applied from right and left.

In **Figure 2**, there is a cross-sectional view of the B type. The swaging fitting is axially and radially pressured to the first pipe from the right and then from the left to the second pipe, resulting in plastic deformation. Pipes and middle part of c fitting exposed to pressure load does not exceed elastic loading. When pipes are of different materials, different swaging parameters must be applied from right and left.

In **Figure 3**, there is a cut view of the Type C. The swaged fitting tube piece is clamped radially to the first tube from the right and then from the left to the second tube simultaneously with two pressure booster wedge rings. In the meantime, sealing is tried to be achieved by creating plastic deformation. Pipes and middle part of circle form fitting exposed to pressure load does not exceed elastic loading. When pipes are of different materials, different swaging parameters from right and left and different riser fitting designs must be applied.

In this study, the swaging circle fitting that will work without leakage under high or low pressures is tightened with each of the three different methods shown under the previous introducing, and the ability to join 2 aluminum alloy pipes is modeled and analyzed by analytical and numerical methods. Different types (Types A, B, and C) designs of aluminum alloy swaged strap fastener were designed in 3D

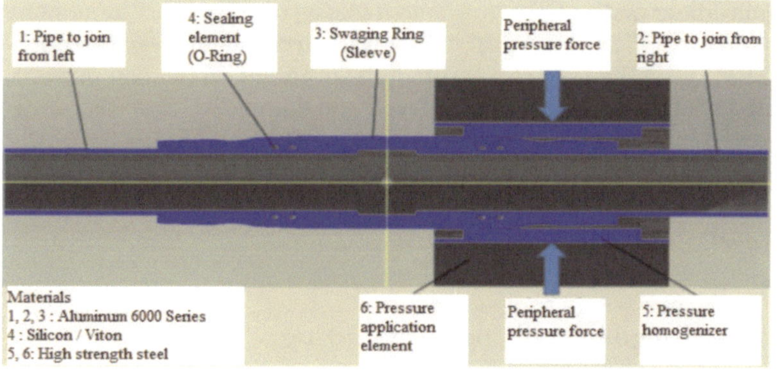

Figure 1.
A type swaging circle fitting; pressure equalizer design features in the radial direction [13].

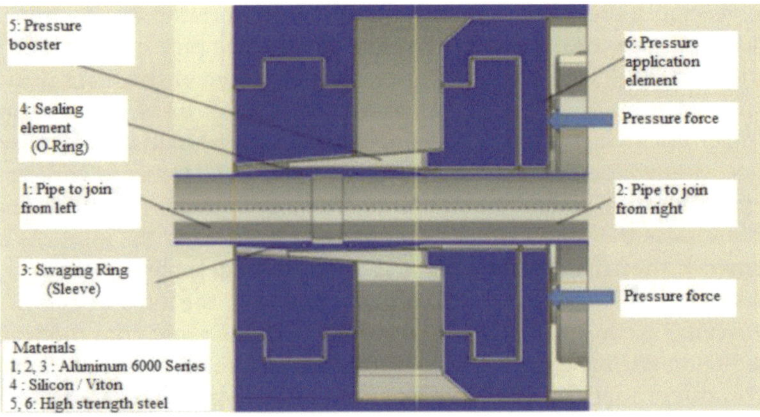

Figure 2.
B type swaging circle; radial swaging designs with pressure against axial tension and pressure booster [13].

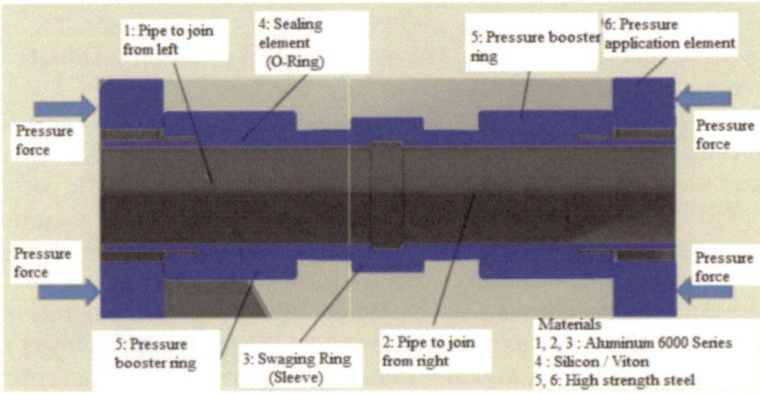

Figure 3.
Swaging type C; radial swaging by tightly compressing the pressure booster (wedge) fitting to the swaging circle fitting by axial pressure force [13].

and mathematically solved using the finite element method. Due to the nature of the numerical method, swaged circle pipe parts and pipes that are connected to each other and may be in different materials are subject to elastic–plastic (bi-linear material definition) deformations. For this reason, the numerical solution could only be realized with a non-linear method [14–16]. In addition, since the material is exposed to plastic pressure and there is friction between the elements, the solution with the non-linear method becomes inevitable again [14, 16]. The nonlinear solution of the mathematical numerical model was carried out using the finite element method (FEM) with the help of a commercial program on the computer. Different swaging circle fitting designs were analyzed and compared, and the connection has been improved by optimizing with swaging circle fitting design changes.

The numerical method proposed in this study are useful for the practicing engineers in the rational design of pipe connections in detail designing and structural analysis.

2. Analytical design of swaging circle fitting

The swaging circle fitting principle basically relies on mechanical-elastic deformation of the fitting with the external pressure load and also subjecting the inner connection pipe to elastic–plastic deformation.

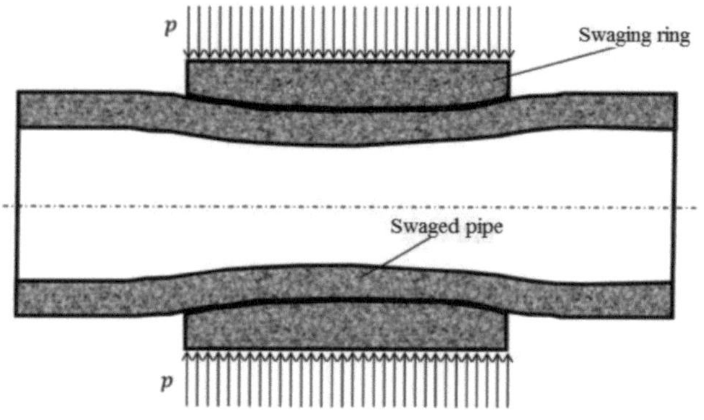

Figure 4.
Swaging principle with swaging circle fitting (sleeve) [13].

Figure 4 shows the principle of elastic–plastic deformation applied to the inner pipe and itself by applying pressure to the swaging circle fitting.

In general, stresses occur in the circumferential and radial directions in a pipe loaded under an externally applied pressure. Its radial stress is equal to the applied pressure on the pressure surface and 0 at the other inner surface. Ratio of pipe inner radius to thickness is in the thin pipe class $r_i/s > 5$, calculations are made according to the middle radius and very small variation in thickness can be ignored [17]. Based on this acceptance, the centrifugal tension equals half the pressure:

$$\sigma_r = \frac{p}{2}; s \ll r_m \tag{1}$$

Here: σ_r [MPa]: centrifugal tension,
p [MPa]: pressure applied,
d_a [mm]: outer diameter of the pipe,
s [mm]: pipe thickness,
$r_m = \frac{d_a - s}{2}$ [mm]: middle radius of the pipe.

Figure 5 shows the circumferential tension σ_{tan} and other parameters occurring under the pressure applied on the pipe. Accordingly, the following equation is written in accordance with the principle of equality of forces in the horizontal direction:

$$\rightarrow \sigma_{tan} = p\frac{r_m}{s} \tag{2}$$

Here: σ_{tan} [MPa]: circumferential tension,
p [MPa]: pressure applied,
D_m [mm]: medium diameter of the pipe,
s [mm]: pipe thickness,
$r_m = \frac{D_m}{2} = \frac{d_a - s}{2}$ [mm]: medium radius of the pipe.

Thus, when the equivalent tension is geometrically collected, it is found as follows:

$$\sigma_E = \sqrt{\sigma_{tan}{}^2 + \sigma_r{}^2 - \sigma_{tan}\sigma_r}$$

$$\sigma_E = \sqrt{\left(p\frac{r_m}{s}\right)^2 + \left(\frac{p}{2}\right)^2 - p\frac{r_m}{s}\frac{p}{2}}$$

$$\sigma_E = p\sqrt{\left(\frac{r_m}{s}\right)^2 + \frac{1}{4} - \frac{r_m}{2s}} \tag{3}$$

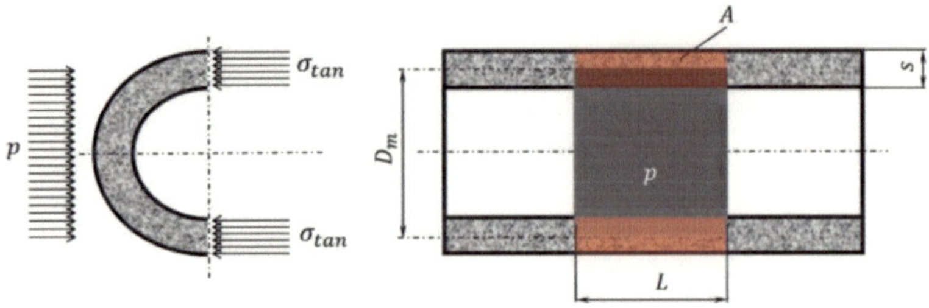

Figure 5.
Peripheral (tangential) stress caused by the pressure pulse [13].

Since the pipe is applied with a minimum degree of yield tension, the least required minimum pressure for this is found by the following formula:

$$p_{min} = \frac{R_{p0,2}}{\sqrt{\left(\frac{r_m}{s}\right)^2 + \frac{1}{4}} - \frac{r_m}{2s}} \tag{4}$$

$$\sigma_{tan} A = pD_m L \rightarrow \sigma_{tan} 2sL = pD_m L$$

$$\rightarrow \sigma_{tan} = \frac{pD_m L}{2sL}$$

Here: p_{min} [MPa]: required minimum pressure,
$R_{p0,2}$ [MPa]: yield tension.
Example: Swaging circle fitting: R-29,4x2–6061-T6;
Yield value: $R_{p0,2} = 240$ MPa.
$\rightarrow r_{m,a} = \frac{d_a - s}{2} = \frac{29,4-2}{2}$ mm $= 13.7$ mm; medium radius of the swaging circle fitting.

$$\rightarrow p_{min,a} = \frac{240 \text{ Mpa}}{\sqrt{\left(\frac{13,7}{2}\right)^2 + \frac{1}{4}} - \frac{13,7}{2 \cdot 2}} = 33,7 \text{ MPa} = 337 \text{ bar}$$

Inner pipe: R-25,4x0,75–6061-T6;
Yield value: $R_{p0,2} = 240$ MPa.
$\rightarrow r_{m,i} = \frac{d_{a,i} - s_i}{2} = \frac{25,4-0,75}{2}$ mm $= 12,325$ mm inner pipe medium radius.

$$\rightarrow p_{min,i} = \frac{240 \; Mpa}{\sqrt{\left(\frac{12,325}{2}\right)^2 + \frac{1}{4}} - \frac{12,325}{2 \cdot 0,75}} = 14,8 \; MPa = 148 \; bar$$

Total required minimum pressure: $p_{min} = p_{min,a} + p_{min,i} = 33,7$ MPa $+$ $14,8$ MPa $= 48,5$ MPa.

In order to create not only elastic but also permanent plastic deflection in compressed pipes, it is recommended to apply a minimum pressure of more than 30% to the swaged flat fitting. Example: $p = 1,3p_{min} = 1,3 \cdot 48,5$ MPa $= 63,0$ MPa.

Listed below are the reasons for the design features marked a to d in **Figure 6** that correspond to the requirements of the ring:

a. In principle, as shown in **Figure 4**, the deformation of the tube will be greater in the middle and decrease parabolic towards both sides of the ring.

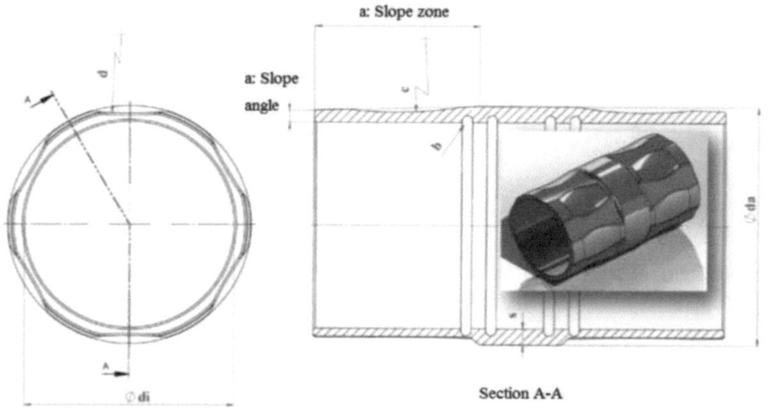

Figure 6.
Swaging circle fitting design [13].

Accordingly, the highest stresses in the inner tube will not be in the middle region, but in the region corresponding to the bottom of the fitting edge. Homogenizing the stress requires homogenizing the plastic deformation, requiring a more convenient beat circle connection. For this, a very small slope (<1.5°) is given on the outer surface of the ring.

b. A circumferential groove is made for the inner diameter of the fitting for 2 sealing elements (O-Rings) in each leakage direction.

c. In the axial direction, a circumferential groove is formed in the middle of the pressure zone of the fitting to make it difficult for the pipe to exit the ring. When the fitting is tightened, a similar plastic deformation is made in the inner pipe, making it difficult for the inner pipe to come out of the fitting in the axial direction.

d. The circumference of the pressure area of the fitting is not circular but rather wavy, making it difficult for the tube to exit the fitting in the circular direction of rotation. When the fitting is tightened, a similar plastic deformation occurs in the inner tube and the inner rod is prevented from rotating in the ring.

The design features of the swaged circle fitting are illustrated in 3D in **Figure 6** and described above. The analyzes of these different design features are analyzed numerically using the finite element method (FEM).

3. Numerical solutions and interpretations of mathematical models

The flat swaged circle fitting described in **Figure 6** and the other 4 swaged circle fittings from A to C are designed in 3D with commercial program (SolidWorks® 2013). The described 3D Models (A, B, and C) are transferred to the numerical commercial solver program (ANSYS Workbench® 14.5) using the Finite Element Method and analyzed by adding physical parameters.

In this study, FEM-Analysis was performed by considering 4 different geometric variations of swaged circle fitting. These have the following geometric design features shown in **Figure 6**:

- Pattern "Flat" swaged circle fitting
- In Model "A", small curved swaged circle fitting
- In "B" model; Small curved swaged circle fitting with inner groove for sealing
- In Model "C", small curved, axially waved swaged circle with inner groove for sealing

The Finite Element Models of 4 different fitting designs is shown in **Figure** 7 (in Mesh form).

Figure 8 shows three separate stepwise loading cases applied in all of the Finite Element Models in Simulation.

The Finite Element Models are solved numerically elastically-plastically by non-linear method. The pictures below show the deformation values of these solutions for the 3 loading cases described above. In particular, the high percentage of plastic deformation and homogeneity indicates the quality of the connection.

In the "Flat" Model in **Figure 9**, since plastic deformation is not lagging behind when the flat swaging circle fitting is released after being subjected to elastic–plastic

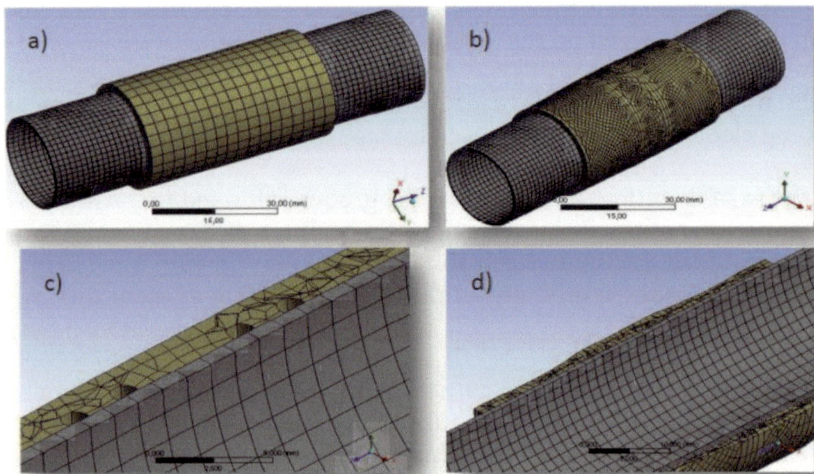

Figure 7.
Finite element model of swaged circle fitting design; a) flat; b) curve "a"; c) curved and fluted "B"; d) curved, ribbed and wavy "C" [13].

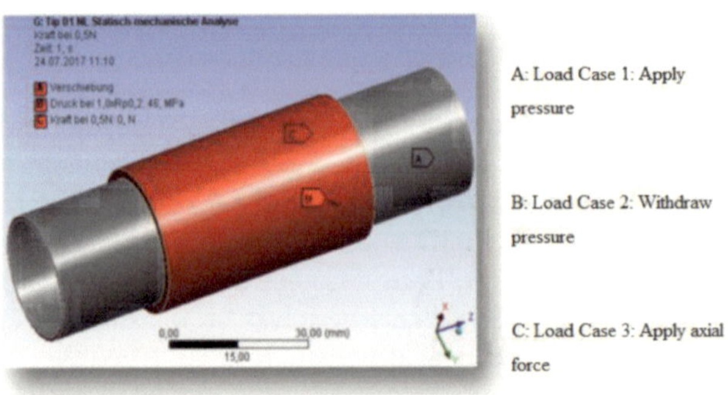

Figure 8.
Applied load cases (LC: Load case) [13].

deformation by applying pressure, it comes out under axial load very easily. Plastic deflection corresponds to approximately 12% of the total deflection. The pressure deformation in the inner pipe is very low.

In Model "A" in **Figure 10**; since plastic deformation is lagging behind when the small curved flat swaging circle fitting is released after being subjected to elastic–plastic deformation by applying pressure, it is now difficult to comes out under axial load. Plastic deflection corresponds to approximately 39% of the total deflection. Plastic deformation shows density in two places and is not homogeneous.

In Model "B" in **Figure 11**; since plastic deformation is lagging behind when the small curved flat swaging circle fitting with internal groove for sealing is released after being subjected to elastic–plastic deformation by applying pressure, it is now

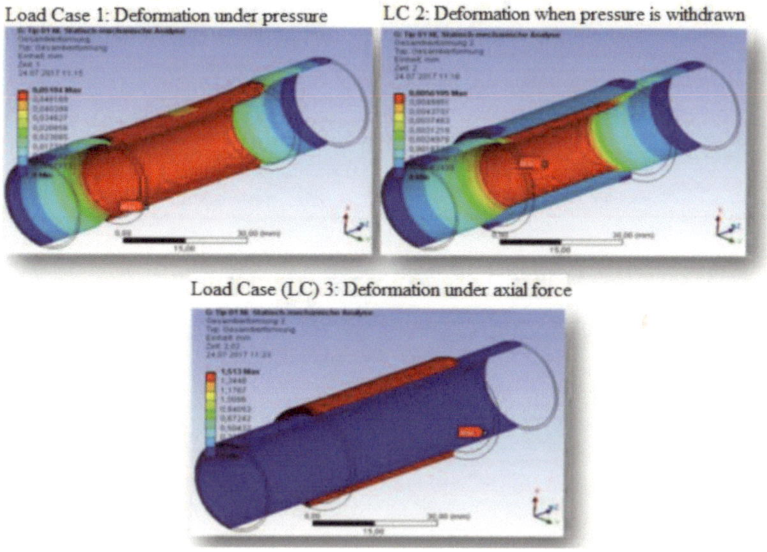

Figure 9.
Elastic–plastic deformations as flat fitting model FEM solution results [13].

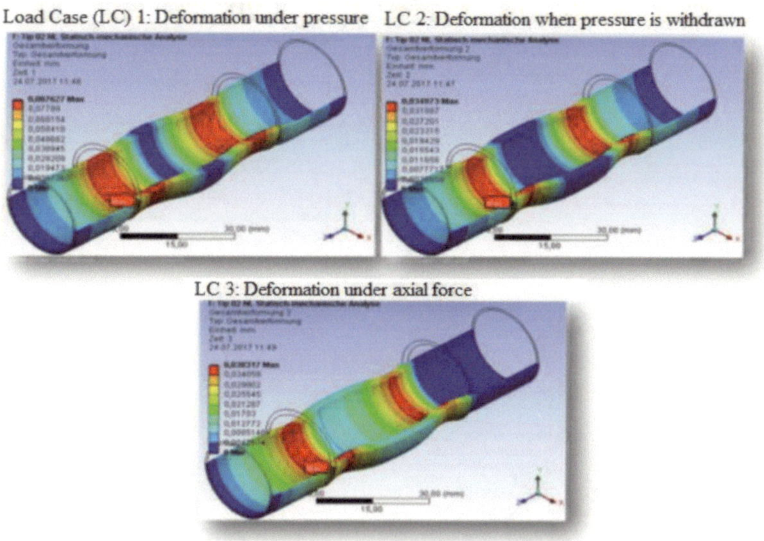

Figure 10.
Elastic–plastic deformations as model "a" FEM solution results [13].

difficult to comes out under axial load. Plastic deflection corresponds to approximately 17% of the total deflection. Plastic deflection distribution is more homogeneous than Model "A" but has a lower percentage.

In Model "C" in **Figure 12**; since plastic deformation is lagging behind when the small curved, axially wavy flat swaging circle fitting with internal groove for sealing is released after being subjected to elastic–plastic deformation by applying pressure, it is now difficult to comes out under axial load. Plastic deflection corresponds to approximately 78% of the total deflection. In this model, plastic deflection is both high and in the most homogeneous state. The deflection in the connected pipes is

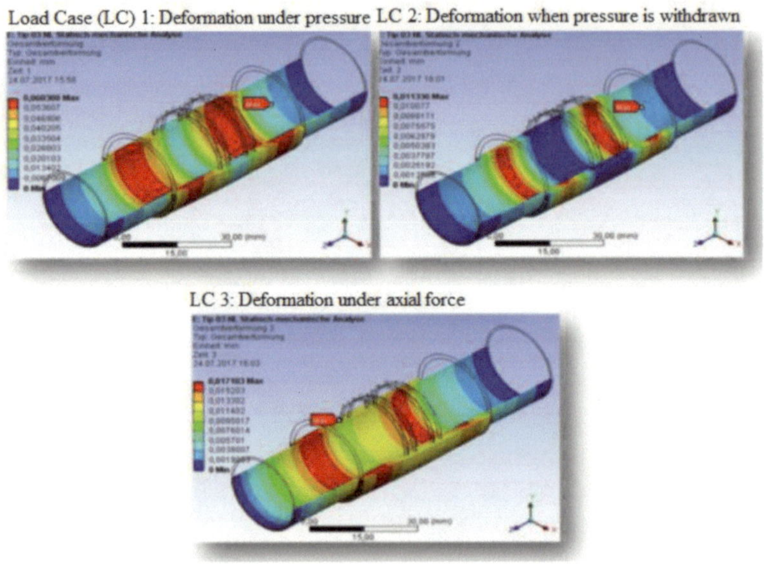

Figure 11.
Elastic–plastic deformations as model "B" FEM solution results [13].

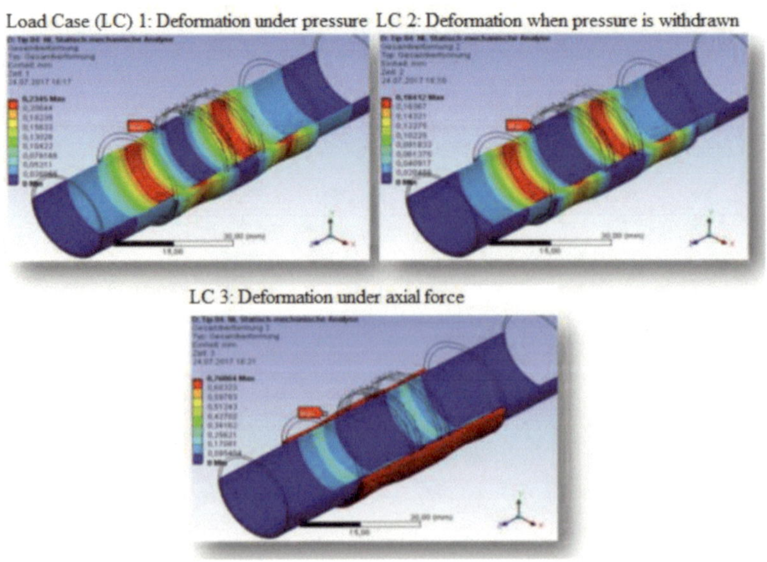

Figure 12.
Elastic–plastic deformations as model "C" FEM solution results [13].

within the elastic limits. It is possible to say that this connection showed a very good solution in terms of both sealing and strength

4. Conclusion

To conclude, in this study, it has been determined that the effect of the swaging circle fitting design on the tightened fittings has a great effect on the swaging force besides the parameters such as material and mechanical properties, use environment, swaging method, swaging circle fitting material, and that the most and homogeneous swaging occurs in the Model type “C” (curved, grooved and wavy) clamping fitting design [13].

Although it is possible to combine different materials and combine two different geometries with different methods, it has been shown in this study that combinations should be modeled and analyzed with advanced engineering since they are exposed to thermal and mechanical loads in terms of usage environments and has high requirements such as sealing [13].

Acknowledgements

I would like to thank my friend Osman Akgun, doctorate student at Gazi University for supporting to this study.

Author details

Ahmet Atak
Gazi University, Ankara, Turkey

*Address all correspondence to: ahmetatak@gazi.edu.tr

IntechOpen

© 2021 The Author(s). Licensee IntechOpen. This chapter is distributed under the terms of the Creative Commons Attribution License (http://creativecommons.org/licenses/by/3.0), which permits unrestricted use, distribution, and reproduction in any medium, provided the original work is properly cited. CC BY

References

[1] Pyplok Fittings Company. A Tube Mac Manufactured Product: Introduction Part. pp. 1-6, Webpage: PAYLOC.COM.

[2] Robert, D. J., K. Soga, and T. D. O'Rourke. Pipelines subjected to fault movement in dry and unsaturated soils. International Journal of Geomechanics 2016;16:5. DOI:10.1061/(asce)gm. 1943-5622.0000548

[3] Karpenko Y. A., Akay A. (2001). A numerical model of friction between rough surfaces. Tribology International 34:531-545.

[4] International Association for Bridge and Structural Engineering (2010). Swaged fittings under tension and fatigue loads. Structural Engineering International 3:284-290.

[5] Adams G. G., Nosonovsky, M. (2000). Contact modeling-forces. Tribology International 33, 431-432.

[6] Mali J. B., Rane S. B., et al. (2016). Modelling and finite element analysis of double ferrule fitting. Journal IJTARME 5(1), 83-87.

[7] Smith B. Alternatives to conventional welded pipe systems. (2016). Business Development Manager at Parker Hannifin Corporation, Home Page of Power Engineering International.

[8] Zhang, Q., Jin, K., Mu, D., et al. (2015). Energy-controlled rotary swaging process for tube workpiece. Int J Adv Manuf. Technol. 80, 2015–2026.

[9] FENN HomeBlog, Swaging machines: What they are and how they work. Available online: https://www.fenn-torin.com/blog/how-swaging-machines-work/ (Accessed on July 2020).

[10] Greenwood J. A. (1997). Analysis of elliptical Herzian contacts. Tribology International 30(3), 235-237.

[11] Mohammed, B., Demagh, R., Derriche, Z. (2020). Structural behavior of pipelines buried in expansive soils under rainfall infiltration (part I: Transverse behavior). Civil engineering journal 6(9), 1822-1838. DOI: 10.28991/cej-2020-03091585

[12] Al-Khazaali, Mohammed, and Sai K. Vanapalli. (2019). A novel experimental technique to investigate soil–pipeline interaction under axial loading in saturated and unsaturated sands. Geotechnical Testing Journal 43(1): 20180059. DOI:10.1520/gtj20180059

[13] Atak, A. (2021). Analytical and numerical model of aluminum alloy swaging ring design to study the effect on the sealing for piping systems. Civil Engineering Journal 7(1), 107-117.

[14] Zhou Sriskandarajah Bamane Dugat S. (2018). Developing Weld Defect Acceptance Criteria for a Swaged Pipe-in-Pipe System. DOI:10.4043/29019-MS. Conference: Offshore Technology Conference.

[15] Mazina Z. R., Seysenov S. Zh., Abyzgildina S. Sh., Tlyasheva R. R. (2020). Ensuring safe operation of the piping connection of apparatus column type. Journal of Physics Conference Series 1515:042041. DOI:10.1088/1742-6596/1515/4/042041

[16] Atak A. (2020). Experimental determination and numerical modeling of the stiffness of a fastener. 62. 12 © Carl Hanser Verlag, München materials testing.

[17] efunda engineering fundamentals. Applications Pressure Vessels. Formula home page (27.07.2020). https://www.efunda.com/formulae/solid_mechanics/mat_mechanics/pressure_vessel.cfm